Partaking of God

*Chapter 1 * * ***

Chapter 10

Job 38
Psalm 104
Psalm 148
Daniel 3: 79-81

Partaking of God

Trinity, Evolution, and Ecology

Denis Edwards

A Michael Glazier Book

LITURGICAL PRESS

Collegeville, Minnesota

www.litpress.org

A Michael Glazier Book published by Liturgical Press

Cover design by Jodi Hendrickson. Cover photo: Dreamstime.

Library of Congress Cataloging-in-Publication Data

Edwards, Denis, 1943–
     Partaking of God : trinity, evolution, and ecology / Denis Edwards.
       pages   cm
     "A Michael Glazier book."
     Includes index.
       ISBN 978-0-8146-8252-4 — ISBN 978-0-8146-8277-7 (ebook)
     1. Human ecology—Religious aspects—Christianity.   2. Evolution (Biology)—
     Religious aspects—Christianity.   3. Creation—History of doctrines.
     4. Nature—History of doctrines.   5. Trinity—History of doctrines.
     6. Darwin, Charles, 1809–1882.   7. Athanasius, Saint, Patriarch of Alexandria,
     –373.   I. Title.
     BT695.5.E55   2014
     231.7'6—dc23   2013049060

Contents

To William R. Stoeger, S.J.
Friend, Mentor, and Colleague

A Personal Beginning

A great deal of preaching and theology—particularly since the Reformation and the Enlightenment—has focused on the relationship between God and the individual human being. In many instances there has also been a focus on the relationships of human beings to one another and to social justice. What has been largely left out is the rest of the natural world. Meanwhile the natural world has been treated ruthlessly, as something for humans to exploit, with little or no regard for consequences, as if it were simply an infinite resource for endless economic growth.

We can no longer ignore the natural world. It is demanding our attention: the climate is changing rapidly, soils are eroding, once fertile land is turning to desert, fisheries are collapsing, rivers are polluted and running dry, forests are shrinking, water tables are falling, glaciers and ice caps are melting, coral reefs are bleaching, the oceans are becoming more acidic and warming, and plant and animal species are suffering extinction at an increasing rate.[1]

Our irresponsible use of fossil fuels, along with other human action, contributes to the climate change that will accelerate the extinction of many species and soon cause extreme suffering to human beings. We are destroying habitats and losing biodiversity. Our actions are a betrayal of the ethical responsibility we have to other forms of life. They also betray future generations of human beings. Taken together, these factors constitute an urgent issue for contemporary theology: first, because for Christian believers they are also a betrayal of God, the giver of creation; and, second, because theology has,

---

[1] This is my reworking of some words of John Surette, SJ, writing to his fellow Jesuits in his "The Dream of an Older Jesuit." http://ecojesuit.com /the-dream-of-an-older-jesuit/4505/.

within its traditions, resources that can inspire a respectful, loving, and responsible stance before other creatures and all that supports life on our planet.

The human community has never had to face anything like the crisis of life that is already upon us early in this twenty-first century. Yet there is a highly visible tendency evident in many everyday attitudes to act as if the crisis does not exist. We see this most clearly in the continuing denial of overwhelming scientific evidence by some political leaders and media commentators. Very importantly, however, there is also good news. There is a growing group of people from every walk of life who are committed to a new way of being on Earth. They see themselves as part of the one community of life, respecting the integrity of the land, the rivers, and the seas and valuing the other species of Earth and their habitats.

We Christians need to be involved alongside others in this movement of ecological conversion. What is asked of us, I believe, is a commitment to it from the depths of our faith in the God of Jesus Christ. If this is so, then it becomes an urgent priority to develop a Christian theology of the natural world for the twenty-first century. The central issue is our understanding of how the rest of the natural world relates to God; closely connected is our view of the relationship of human beings to the rest of the natural world before God.

For a Christian believer, the kind of theology required needs to be more than simply an attempt to show the connection between creatures and God the Creator. The conviction that God is the Creator of the universe as well as the Earth and all its creatures is certainly central to Christian faith. It is, however, part of a much larger picture—one of a God who creates, who gives God's very self to the creation in the incarnation of the Word, and who brings healing and fulfillment to creation. So the question becomes: How is the natural world related to the God of creation, incarnation, and final fulfillment?

It is impossible to talk with integrity about the natural world today without attending to its evolutionary character. The science of the last two centuries has provided us with two huge developments in the understanding of the evolutionary nature of the world of which we are a part. The first is Charles Darwin's nineteenth-century discovery of the evolution of life by means of natural selection and its confirmation by the biological sciences of the twentieth and the twenty-first centuries. Evolutionary biology offers us a convincing evolutionary explanation for the emergence of all the diverse forms of life on Earth back to their microbial origins about 3.7 billion years ago. The second is the twentieth-century discovery based originally on Albert Einstein's theory of general relativity and Edwin Hubble's astronomical observations that we do not belong to a static universe that is limited to the Milky Way Galaxy we see around us. Rather, the observable universe is made up of more than a hundred billion galaxies that have been expanding and evolving dynamically since the universe's origin about 13.75 billion years ago—the big bang theory.

A theology of the natural world must involve the best understanding of the nature of the universe and of life that science offers. It must come to terms with evolution. And if it is to draw on what is central to Christianity, it must involve the theology not only of creation but also of God's saving action in Jesus Christ. It will need to spring from the fullness of Christian faith, God's self-giving to creation in Christ and in the Spirit. This means it necessarily involves the Christian theology of the Trinity.

Why Another Book?

Some of my friends raise their eyebrows when I tell them that I am writing a book on the Trinity in relationship to the natural world. They rightly point out that I have done this already, most explicitly in *The God of Evolution: A Trinitarian Theology* (New York: Paulist Press, 1999) but also in other publications. Why do it again?

My answer is that I now find myself not fully satisfied with the theology of the Trinity that I outlined in the earlier book. It is primarily a theology of divine Communion, influenced by contemporary writers like John Zizioulas and particularly Walter Kasper, as well as the medieval theologian Richard of St. Victor. It builds on the biblical theology of mutual indwelling found in the Supper discourse of the Gospel of John (13:31–17:26). It is not that I now reject this theology—far from it. I embrace it and find it essential to a theology of the Trinity for today. But I think more is needed. I do not wish to replace such a Communion theology but to complement it and show better its grounding in the biblical narrative of God's creating and saving action. I now think what is needed is a more dynamic account of the Trinity in action—particularly for a theology of the natural world that has an emergent, evolutionary character.

All our theology is limited and partial. The experience of theologians who teach courses on God the Trinity year after year is that God remains an incomprehensible, unspeakable, but most beautiful Mystery of love. Before this Mystery of love all theologies fail. Yet we must speak about God. In searching for ways to express what escapes us we are always being drawn into the new. And this new often comes from a deeper entry into the great theologies of the past. In the regular rhythm of teaching, there is the chance to focus one year on a deeper reading into Aquinas, another year on Augustine, and yet another on Basil of Caesarea or Gregory of Nazianzus.

In this way I have found myself drawn in recent years to focus in a new way on Athanasius of Alexandria. It is his insights into God that I seek to bring into a creative dialogue with an evolutionary world in this book. What draws me to Athanasius is that in his work the theology of the Trinity seems still young. It is a theology in the process of development. It is a dynamic theology, filled, I believe, with the energy of the Spirit. It is always radically and refreshingly biblical. It is always about God's action in creation and salvation. It is a theology concerned with the transformation of creatures in God that he calls deification. His is a theology of the Trinity in action, of God acting through the Word and in the Spirit in creation and in deification.

So what is new in this book is the attempt to ground a theology of the natural world in the dynamic trinitarian theology of Athanasius. What else is new is the attempt to build on, and to develop, Athanasius's theology in the light of evolution so that it becomes articulated as a theology of the Word of God as the divine Attractor and the Holy Spirit as the Energy of Love in evolutionary emergence. Writing this book has provided me with the chance to think again, hopefully more deeply, about four issues that confront this kind of theology: the question of God's suffering with creatures in the light of the costs of evolution; the idea of the humility of God in relation to the relative autonomy of evolutionary processes; church teaching on the human soul in relation to the insights of neuroscience into the mind and the brain; and the doctrines of grace and original sin in relationship to evolution.

This book is divided into three parts. The first part is made up of three chapters that explore Athanasius's trinitarian and incarnational theology as a foundation for a theology of the natural world. The second part contains five chapters that address key theological developments that I see as required for such a theology in today's evolutionary context. The third part consists of two chapters that come back to the beginning, bringing the theology of God developed throughout the book into explicit engagement with the ecological crisis we face.

Beyond Male and Female

For many Christians today, the exclusive use of Father-Son language in talk of the Trinity and the constant use of the male personal pronoun for each of the divine persons constitute a serious problem. This language can have the effect of undermining the truth that women as well as men are made in the image of God—and thus function to reinforce inequality. And when the eternal divine persons are taken to be literally male, the theology of God is damaged and distorted.

The theologians of the church have understood that Father-Son language is analogical. Gregory of Nyssa, for example, in his *Commentary on the Song of Songs*, writes that when the words father and mother are applied to God, "Both terms mean the same, because there is neither male nor female in God."[2] Richard of Saint Victor is equally explicit.[3] While theologians may always have known that

---

[2] Gregory of Nyssa, "Homily Seven," in *Saint Gregory of Nyssa: Commentary on the Song of Songs*, trans. Casimir McCambley (Brookline, MA: Hellenic College Press, 1987), 145.

[3] "We must observe that there are two [different] sexes in the human nature, and for this reason the terms denoting relationship are different according to the

there is neither male nor female in the Trinity, and that human beings, male and female, are both made in the image of God (Gen 1:27), this knowledge has been unavailable to the great majority of Christians. Elizabeth Johnson, along with other Christian feminist theologians, has begun to address this urgent issue, and to open up new pathways in our speech about the triune God, particularly through the use of Wisdom categories.[4]

In this book I will use Father-Son language when I am following Athanasius, and when I need to do so for the sake of clarity, but I will not use male personal pronouns of God. When I can, I will use "God" for the first person, following much of the Bible, particularly Paul, and I will also use Source of All. When it seems appropriate, I will highlight Athanasius's use of Wisdom/Word language and focus attention on the range of analogies he uses as alternatives to Father-Son, such as God and God's Word, God and God's Wisdom, Light and its eternal Radiance, and Spring and its River of living water.

When speaking in my own voice, I will point to the equivalence of fathering and mothering images and language in speech about the Trinity. To better bring this out I will sometimes speak of God as Mother and sometimes as Mother/Father. We need images of God that are taken from human life alongside others from nature, and among the human images, parenting images have a special place. Jesus' use of *Abba*/Father for God, and his invitation to his followers to do the same, is a precious and liberating part of our biblical and Christian heritage (Mark 14:36; Gal 4:6; Rom 8:15), as is his image of the father filled with compassion for his wayward son, running to meet him, throwing his arms around him, and kissing him (Luke

---

differences of the sexes. We call one who is a parent either 'father' or 'mother' [according to their] sex. In case of progeny, [we say] in one case 'sons' and in another 'daughters.' In the divine nature, instead, as we all know, there is absolutely no sex." Reflecting his male-centered culture, Richard goes on to say that it is a convenient custom to use language associated with "the more worthy sex" of "the most worthy being in the universe." *Richard of Saint Victor: On the Trinity* 6.4., trans. Ruben Angelici (Eugene, OR: Cascade Books, 2011), 207.

[4] Elizabeth A. Johnson, *She Who Is: The Mystery of God in Feminist Theological Discourse* (New York: Crossroad, 1992).

15:20). Taken in the context of the Kingdom message of Jesus, this *Abba*/Father language does not reinforce patriarchal domination, but challenges it with a vision of human beings as beloved daughters and sons of a God of limitless tender mercy and love.

This same message is proclaimed in a different way with the biblical image of God as a mother. Like a mother God has carried us from birth, borne us from the womb (Isa 46:3-4). Like a mother God will never abandon us: "Can a woman forget her nursing child, or show no compassion for the child of her womb? Even if these forget, yet I shall not forget you" (Isa 49:14-15). In times of prayer we are encouraged to adopt the attitude to God of the psalmist: "But I have calmed and quieted my soul, like a weaned child with its mother" (Ps 131:2). In an especially beautiful expression, close in some ways to the tenderness of Jesus' *Abba* image and teaching, God speaks these words at the end of Isaiah: "As a mother comforts her child, so will I comfort you" (Isa 66:13).

Maternal and paternal images, as well as those from the natural world, can all speak to us of God. As I will point out, if they are to function in a trinitarian way, they need to be correlational, expressing the dynamics of the relations of origins—the generation of the Word and the procession of the Spirit from the Source of All. We need to face constantly both the limited nature and the dangers of our human language for God. God is a mystery of love that radically transcends all finite human categories and language. When we use male or female language to evoke the divine persons, we do so with the reservation that the divine persons of the Trinity cannot be limited to male or female, but beautifully encapsulate and infinitely transcend both.

I owe a particular debt of gratitude to those who have read these chapters and given me very helpful responses and suggestions, particularly John Haughey, SJ, and Patricia Fox, RSM, James McEvoy, Alastair Blake, Rosemary Hocking, and, for the earlier chapters, Brian McDermott, SJ, and Julie Clague. I am most grateful for the opportunity I had to work on this book at Georgetown University as a Woodstock International Visiting Fellow from September 2012 until February 2013. During this time, I had the blessing of being welcomed to live in the Woodstock house of the Jesuit community at

Georgetown. It was a precious gift to be able to share research, prayer, and evening meals in this household of Jesuit scholars, Gasper LoBiondo, SJ, John Haughey, SJ, Leon Hooper, SJ, Thomas J. Reese, SJ, Thomas Michel, SJ, Thomas Gaunt, SJ, and Dan Madigan, SJ. During this time I benefitted greatly from the chance to discuss the issues of this book with colleagues Elizabeth Johnson, CSJ, of Fordham University, and Mary Catherine Hilkert, OP, of Notre Dame.

I am grateful to Dr. William F. Storrar and to Professor Celia Deane-Drummond for their invitation to lead a seminar with the *Inquiry on Evolution & Human Nature*, at the Center of Theological Inquiry, Princeton University, on November 8, 2013, and to Celia Deane-Drummond for helpful conversations and suggestions for reading in the area of the evolution of cooperation. I am also grateful to Professor Manuel Doncel, SJ, for his invitation to go to Barcelona at the end of February 2013 to offer a lecture to the Faculty of Theology of Catalonia and to engage with the Seminar of Theology and the Sciences. I benefitted greatly from many helpful conversations with him and from his wonderful hospitality and that of the Jesuit community at the Center Borja in Sant Cugat. Both the Princeton Inquiry and the event in Barcelona were generously supported by the John Templeton Foundation.

My research for this book owes much to Professor Ernst Conradie of the University of the Western Cape, South Africa, who ably and generously led an international research program in ecological theology from 2007 to its culmination in a conference near Cape Town in August 2012.[5] I am also much indebted to Niels Gregersen of the University of Copenhagen who, with Mary Ann Meyers of the John

---

[5] My publications in this project form the background for this book: "God's Redeeming Act: Deifying Transformation," *Worldviews* 14 (2010): 243–57; "Athanasius: The Word of God in Creation and Salvation," in *Creation and Salvation Volume 1: A Mosaic of Essays on Selected Classic Christian Theologians*, ed. E. Conradie (Berlin: LIT Verlag, 2011), 37–51; "The Attractor and the Energy of Love: Trinity in Evolutionary and Ecological Context," *The Ecumenical Review* 65, no. 1 (March 2013), 129–44; "Where on Earth Is the Triune God," in *Christian Faith and the Earth*, ed. E. Conradie et al. (London: T & T Clark, forthcoming).

Templeton Foundation, hosted a colloquium in Helsingor, Denmark (August 25–27, 2011) on: "Is God Incarnate in All That Is?"[6] I am very grateful to Hans Christoffersen, Lauren L. Murphy, and all at Liturgical Press, particularly to my copy editor Patrick McGowan. Scripture quotations, except where otherwise noted, are taken from the New Revised Standard Version Bible.

[6] My contribution, "Incarnation and the Natural World: Explorations in the Tradition of Athanasius" is related closely to chapter 3 of this book. The original is to be published in *Incarnation and the Depth of Reality*, ed. Niels Gregersen, forthcoming from Fortress, and is adapted here with kind permission of Fortress Press.

Part One

Athanasius of Alexandria as Resource

A central proposal of this book is that Athanasius of Alexandria can be a helpful resource for a trinitarian theology of creation and salvation that is meaningful in today's evolutionary and ecological context. Before taking up some key aspects of his theology, I will introduce him briefly, outline why I see him as a resource in this new context, and then offer some observations on my approach to interpreting his work.

Introducing Athanasius (c. 295–373)

Athanasius grew up in the great port city of Alexandria, an administrative center for the Roman Empire and a vital source of grain, manufactured goods, and general trade. It was a city known for its learning and culture, a place of Hellenistic, Jewish, and Christian scholarship. Along with Jerusalem, Antioch, Rome, and, later, Constantinople, it was a pastoral and theological center of the Christian church.

Little is known of Athanasius's early life. According to one account, as a young man he was attached to the household of Alexander, the bishop of Alexandria, where he was educated and learned the Scriptures.[1] His later writings give evidence of a wide and deep

---

[1] For Athanasius's life see David Gwynn, *Athanasius of Alexandria: Bishop, Theologian, Ascetic, Father* (Oxford, UK: Oxford University Press, 2012).

knowledge of the Bible. He gives the impression of having scriptural texts ever present to his mind, of having long meditated on them and of knowing them by heart.

Athanasius's life and theology would be deeply intertwined with Arius, an influential priest of Alexandria. Arius had denied the eternity of the Word of God. For him the Word seemed to be a created intermediary between God and the world of creatures. This led to a fierce controversy with his bishop, Alexander, which soon spread beyond Alexandria to other churches. When the emperor Constantine intervened in this dispute and called a council at Nicaea in 325, Athanasius accompanied Alexander to the gathering of bishops as a deacon and a secretary. The council completely rejected the views of Arius and taught that the Word of God is fully and eternally divine, possessing the same substance (*homoousios*) with the Father.

Three years after Nicaea, Athanasius became bishop of Alexandria at about the age of thirty. He inherited a church in serious conflict, not only with Arius and his supporters, but also with the Melitians, a Christian group that had set up its own bishops and rejected Athanasius's authority. For much of his life Athanasius was embroiled in controversy over the full divinity of the Word, not only with other bishops and teachers, but also, in varying ways, with the imperial authorities.

Throughout the East he became the leading defender of pro-Nicene theology and a vigorous opponent of those who held anti-Nicene views, whom he called collectively "Arians." Lewis Ayres describes Athanasius as constructing "Arianism," by drawing his anti-Nicene enemies into an account that presents them as perpetuating a theology that stems from the already condemned Arius. His opponents included not only Arius but also Eusebius of Nicomedia and Asterius, whom he called the "standard-bearer" of Arianism.[2]

Athanasius was bishop of Alexandria for forty-six years, but because of imperial opposition he was exiled five times for a total of seventeen years. During his exiles in Trier, Rome, and in the desert monasteries of Egypt, he formed alliances and wrote impor-

---

[2] Lewis Ayres, *Nicaea and its Legacy: An Approach to Fourth-Century Trinitarian Theology* (Oxford, UK: Oxford University Press, 2004), 107.

tant works. He gave support and leadership to those women and men who were participating in emerging forms of monastic life and helped shape monastic spirituality in Egypt. At key moments the monks of Egypt would be strong supporters of Athanasius. No doubt Athanasius also learnt much from his engagement with them, as seems evident in his *Life of Anthony*. But Athanasius was always a pastoral bishop, highly committed not only to those who took up monastic forms of life but also to all those in his congregations who were living Christian life in the world. He constantly put before all of them the call to follow Jesus, seeing them all as called to the ascetical life.

Athanasius was a powerful advocate for Nicaea's teaching that the Word who becomes flesh in Jesus of Nazareth is fully and eternally divine. Later he would take up the defense of the full divinity of the Holy Spirit. By the end of his life, the Christian community was coming to a new consensus on pro-Nicene theology, and Athanasius's work was being carried forward by other bishop-theologians, particularly the three Cappadocians, Basil of Caesarea, Gregory of Nazianzus, and Gregory of Nyssa. It would find definitive expression in the Council of Constantinople of 381, which reaffirmed Nicaea and taught the full divinity of the Holy Spirit.

Athanasius's most systematic work is his double treatise *Against the Greeks—On the Incarnation*, probably written between 328 and 333, in the period after he became a bishop and before his first exile.[3] Athanasius's later works were more explicitly polemical, particularly his *Orations against the Arians*, the first two of which he wrote in Rome during his second exile about 339–40. Around 357, during his third exile in the monastic communities of the Egyptian desert (356–62), Athanasius wrote to his friend Bishop Serapion of Thmuis, defending the full divinity of the Holy Spirit. This constitutes the first developed theology of the Holy Spirit in the Christian tradition.

Athanasius lived in turbulent and violent times. He was accused by his enemies of violence, but there is no evidence that he himself either used or condoned it. The same cannot be said of some of his

---

[3] See Khaled Anatolios, *Retrieving Nicaea: The Development and Meaning of Trinitarian Doctrine* (Grand Rapids, MI: Baker Academic, 2011), 101.

supporters. Athanasius did use polemical, harsh, and dismissive language of his opponents and their positions. As Khaled Anatolios notes, however, in reading the work of a fourth-century Egyptian bishop, it is important to remember that one is dealing with a culture that is totally other to one's own. What is unacceptable in the twenty-first century may have seemed more like steadfast attachment to the truth in Athanasius's own circle. From Athanasius's own perspective, Anatolios suggests, he would have seen himself as a "persecuted shepherd of an embattled flock who is not only at pains to provide his people with cogent and persuasive reasons for denying 'Arian' doctrine, but who is also quite desperate to coach them in the affective repugnance they ought to feel for such 'blasphemy.'"[4]

The urgency of Athanasius's language comes from his overriding conviction that the denial of the divinity of the Word deconstructs the whole of Christian faith. What finally matters for Christian theology is not Athanasius's personality, which cannot be wholly reconstructed, but his central conviction that Christian faith stands or falls with the full divinity of the Word.[5] He defended this conviction when the power of the empire and much of the church opposed him. He won the support of his own people and the growing monastic community of Egypt and ultimately that of the wider church.

Why Athanasius?

Why turn to Athanasius for a basis for a contemporary theology of the Trinity, one that can engage with twenty-first-century evolutionary thought and the ecological crisis we face? I have already briefly discussed why I find him helpful, but I will now spell out my reasons more fully:

1. *Athanasius sees the Trinity in biblical and narrative terms, in relationship to the divine "economy" of creation and salvation.* In recent times theologians have become aware of the dangers of theologies of the Trinity that are overly abstract and speculative,

---

[4] Khaled Anatolios, *Athanasius* (London: Routledge, 2004), 36.
[5] Ibid., 39.

too remote from their origin in the biblical story of salvation and the lives of believers. Athanasius provides a helpful alternative. The center of attention for Athanasius is the triune God's engagement with creatures in creation and salvation. It is the biblical story he tells. His focus is not on a discussion of God in God's self (the immanent Trinity) in the abstract, but on the true nature of the God who acts in creation and salvation (the economic Trinity). He is intensely interested in who it is that creates and saves, hence his emphasis on the full and eternal divinity of the Word and the Holy Spirit. It is this biblical and narrative approach that I see as providing an appropriate basis on which to build an evolutionary theology.

2. *For Athanasius, the theology of the Trinity is not seen as an aspect of theology, or as one subject within theology, but as the whole of Christian existence.* Unlike many more recent theologies, Athanasius's theology of God the Trinity is a coherent understanding of the entirety of Christian existence in the world.[6] As I hope to make clear, it is also a coherent Christian account of God's relationship not only with human beings but also with the rest of the natural world. It is concerned with God and the whole universe of creatures in God.

3. *For Athanasius, creation and the saving incarnation are deeply and intrinsically interconnected.* Athanasius's theology is built around the identity between the Word who is made flesh, who is the Word of the cross, and the eternal Word in whom all things are created. The Word of creation is the Word of salvation. While Athanasius does not explicitly develop a theology like that of Maximus the Confessor (580–662) in the East or Duns Scotus (1266–1308) in the West, which understands the incarnation as eternally God's intention in the creation of a world of creatures, his thought already interconnects creation and incarnation in a profound way. I will suggest that his theology of the incarnation and the ontological change it brings can be understood as "deep incarnation" in an evolutionary and ecological context.

[6] Anatolios, *Retrieving Nicaea*, 1, 7–8, 11.

4. *Salvation is understood in a theology of deification.* Along with many other Christians, I believe that we need to rethink our theology of salvation in Christ, and I find Athanasius's theology of deification as one that offers a fruitful resource. Athanasius has a deep grasp of the evil of human sin and the need for salvation. He makes use of various biblical images and concepts for redemption in Christ. But he also contributes to an overarching theological vision of Christ's saving work as deifying, as an ontological change in reality by which human beings are transformed by the grace of the Spirit so that they are adopted as daughters and sons in Christ and participate in the divine life of the Trinity.

5. *Athanasius's view of deification involves not just human beings but in some way the whole natural world in transformation and fulfillment in Christ.* Athanasius sees the incarnation of the Word of God as bringing about a transfiguration of the whole of created reality. The Word of God embraces bodily existence in a world of creatures and transforms this world from within. God becomes a creature of matter and flesh in order that creatures might reach their fulfillment in God. While his primary focus is not on the natural world, but on human beings, Athanasius includes the wider creation in his view of salvation and deification.

Interpreting Athanasius in a New Moment

I will seek in this volume to appropriate Athanasius's theology anew in the context of the evolutionary and ecological issues we face. To this end, I will engage directly with Athanasius's views in the first section of the book, but will not attempt a full presentation of his theology. In recent times, historical theologians have offered several such overviews and I have learned a great deal from them.[7]

---

[7] See Khaled Anatolios, *Athanasius: The Coherence of His Thought* (New York: Routledge, 1998); Anatolios, *Athanasius*; Anatolios, *Retrieving Nicaea*; John Behr, *The Nicene Faith: Part 1: True God of True God, The Formation of Christian*

My approach and my emphasis will be partial and selective. I will do my best to respect Athanasius and his own theological context while reading him from a context—evolutionary and ecological—very different from his own. My proposal is that his classical theology can offer the basis for a theological vision that has new meaning in today's context.

In the second section of this book I will suggest ways in which I see the need to explore new developments in a trinitarian theology of creation for today, developments not envisaged by Athanasius. Of course, I do not claim Athanasius's authority for these theological developments. Athanasius is not a contemporary evolutionary or ecological theologian, but a fourth-century defender of pro-Nicene theology against its opponents. The developments proposed in the second half of this book, then, can be warranted only by their fidelity to the liberating good news of the Christian tradition and by their plausibility in an evolutionary and ecological context.

Theology, vol. 2 (Crestwood, NY: St. Vladimir's Seminary Press, 2004); Gwynn, *Athanasius of Alexandria*; Peter Leithard, *Athanasius* (Grand Rapids, MI: Baker Academic, 2011); Alvyn Pettersen, *Athanasius* (London: Geoffrey Chapman, 1995); Thomas Weinandy, *Athanasius: A Theological Introduction* (Aldershot, Hampshire: Ashgate, 2007). For more general background, see Lewis Ayres, *Nicaea and its Legacy*; R. P. C. Hanson, *The Search for the Christian Doctrine of God: The Arian Controversy, 318–381* (Edinburgh: T&T Clark, 1988); Rowan Williams, *Arius: Heresy and Tradition* (London: Darton, Longman & Todd, 1987).

Chapter 1

Trinity in Act
Creating

We live in a time when the sciences offer us awe-inspiring insights into the nature of the universe, the history of the evolution of life on Earth, and the functioning of the human brain. These new insights mean that we think about ourselves and the world we inhabit in very different ways from earlier generations. At the same time, we are confronted by the terrifying fact that we human beings are abusing the natural world on a massive scale, doing irreparable damage to other species and to the life-support systems of the planet.

These two realities call for a response from those of us who see the world in Christian terms. There is a clear need to think again about how we are to understand the natural world and our part in it in relationship to God. This will require digging deeply into the tradition to find resources for a Christian theology of creation in this new context. We need a renewed theology of the natural world, one that includes and embraces the community of life on Earth, understood in evolutionary terms and as under extreme threat.

The proposal I am making is that Athanasius's fully trinitarian theology of creation and deification can be an important resource for this new moment. His theology tells of the God who acts in history, the God of creation and of the incarnation that brings about the deification of creatures. It is a theology of the Source of All acting through the Word and in the Holy Spirit, in both creation and new creation: "The Father creates and renews all things through the Son and in the Holy Spirit."[1] Athanasius's interest is in the biblical

---

[1] Athanasius of Alexandria, *Letters to Serapion on the Holy Spirit* (= *Ep. Ser.*), 1.24, in Khaled Anatolios, *Athanasius* (London: Routledge, 2004), 225.

story of God's action for us and for all creatures. At the center of the narrative is the incarnation of the Word, which culminates in Jesus' death and resurrection that overcomes death and transforms creaturely existence. What Athanasius seeks to defend above all is the utter realism of this story, the conviction that Jesus is in all truth the eternal, divine Word and Wisdom of God, and that the Spirit who unites creatures to this eternal Word is not a creature but the uncreated gift, the Holy Spirit of God.

It is important to note that Athanasius discusses creation from a very particular perspective, that of the cross of Jesus. When he treats of the creation of the world in *Against the Greeks* and of humanity in *On the Incarnation*, his starting point is the scandal of the cross. Both books are written as a defense of the cross against those who mock the idea of a crucified savior.[2] Athanasius's central strategy against those who scorn the cross is to show that the one on the cross is the divine Word of God, who, by entering into death, brings salvation to the whole creation. Those who slander the cross, he says, fail to understand that the crucified Christ is "the Saviour of the universe and that the cross was not the ruin but the salvation of creation."[3] For Athanasius the one on the cross is the true and eternal Word and Wisdom of God through whom all things are created.[4]

In this chapter, I will focus on four ideas from Athanasius's theology of creation that I believe have a great deal of meaning in today's evolutionary and ecological context: (1) God continually creates all things through the Word and in the Spirit; (2) The fruitfulness of the universe of creatures springs from the eternal

---

[2] See Khaled Anatolios, *Athanasius: The Coherence of His Thought* (London: Routledge, 1998), 28; John Behr, *The Nicene Faith: Part 1: True God of True God*, The Formation of Christian Theology, vol. 2 (Crestwood, NY: St. Vladimir's Seminary Press, 2004), 171.

[3] Athanasius of Alexandria, "Against the Greeks" (= *C. gent.*), 1 in *Athanasius: Contra Gentes and De Incarnatione*, trans. Robert Thomson (Oxford, UK: Clarendon Press, 1971), 5.

[4] John Behr writes: "It is the Word of the Cross, or the Word on the Cross, that Athanasius expounds by describing how all things have come into being by and for him; it is Christ himself that Athanasius is reflecting on, not the creation accounts in and of themselves." Behr, *The Nicene Faith: Part 1*, 181–82.

generativity of the triune life; (3) The creaturely world exists within the relations of mutual delight of the divine persons; (4) In the relationship of creation, each divine person is immediately present to the creature.

God Continually Creates All Entities through the Word and in the Spirit

Athanasius's strong conviction that the Word of the incarnation and cross is the very same Word through whom God creates all things springs from his reading of key biblical texts that speak of the Word or Wisdom of God. He understands the expressions "Word of God" and "Wisdom of God" as referring to the same divine reality. He interprets the various texts that refer to the Word of God or the Wisdom of God in an intertextual way, and he understands them in the light of Christ. So, for example, when in First Corinthians Paul calls Christ "the Wisdom of God" (1 Cor 1:24, 30), Athanasius reads this in conjunction with Wisdom's description of her cosmic role in creation in Proverb 8: "When he marked out the foundations of the earth, then I was beside him, like a master worker; and I was daily his delight, rejoicing before him always, rejoicing in his inhabited world and delighting in the human race" (Prov 8:29-31).

A fundamental text for Athanasius is found in the prologue of John's gospel which sings of the Word's role in creation—"All things came into being through him, and without him not one thing came into being" (John 1:3)—and then proclaims of this creative Word, "And the Word became flesh" (John 1:14). Athanasius refers often to the great creation hymn to Christ found in Colossians: "He is the image of the invisible God, the firstborn of all creation; for in him all things in heaven and on earth were created, things visible and invisible, whether thrones or dominions or rulers or powers—all things have been created through him and for him" (Col 1:15-20). Another favorite is the opening of Hebrews, where Christ is seen not only as the Son of God but also as the Radiance of God: "In these last days he has spoken to us by a Son, whom he appointed heir of all things, through whom he also created the worlds. He is the radiance of God's glory and the exact imprint of God's very being, and he sustains all things by his powerful word" (Heb 1:2-3).[5]

Based on these and other texts, Athanasius speaks freely and often of Christ as the Wisdom of God, the Word of God, the Image of God, the Radiance of God, as well as the Son of God. He is not at all unique in the fourth century in seeing the biblical texts referred to above as revealing that God creates through the Wisdom/Word of God. His opponents would in fact be in strong agreement with him. What makes Athanasius's position distinct is his deep conviction that this Wisdom/Word has the very being of God and is God.

Creation: Partaking of the Word

Building on Irenaeus and other early Christian theologians, Athanasius contributes to the development of a robust Christian theology of creation out of nothing (*creatio ex nihilo*). Creatures, he argues, have in themselves absolutely no reason for their existence. They exist only through the sheer divine goodness and benevolence by which the God "beyond all being and human thought" creates a universe of creatures through the Word.[6] Precisely as beyond all creatures, God is able to be boundlessly generous and unthinkably close, and so confer existence on each entity. Because God is beyond all creaturely limits, God can be lovingly and intimately present to created entities.[7] Creation is the relationship whereby the God, who is "by nature invisible and incomprehensible," becomes the generous source of finite created beings.[8]

According to Athanasius, divine generosity characterizes God's relations with creatures from the beginning. Because all creatures exist out of nothing at every point, Athanasius sees them as inherently

---

[5] My translation. While the NRSV translates the Greek word *apaugasma* as "reflection," I have translated it as "radiance," because of its importance as an image in Athanasius's trinitarian theology, where the Father is the Light and the eternal Word is the Radiance of the Light. See also Wisdom of Solomon 7:26.

[6] *C. gent.* 2 (Thomson, *Athanasius,* 7).

[7] Khaled Anatolios speaks of a "dialectical simultaneity" in Athanasius's thought between God's radical "beyondness" on the one hand and God's generous "*philanthrōpia*" on the other, in his *Athanasius,* 40.

[8] *C. gent.* 35 (Thomson, *Athanasius,* 95).

unstable. By themselves they have no hold on existence—it comes as a gift from their participation in the Word of God. Alvyn Pettersen says that, for Athanasius, "creation is both very fragile, but most wonderful." While creatures are fragile in themselves, they are also sublime because they participate in the Word of God. In a sentence that I think captures Athanasius's creation theology perfectly, Pettersen says: "Creation is from the world's side a continuous receiving of God who gives all that it is and has."[9]

Every creature exists because it continually receives its existence from the Word of God. It exists by partaking of the Word. It is not just that creation is originally brought into existence through the Word but also that each creature continues to exist at every moment only by this continuous participation in the Word:

> After making everything by his own eternal Word and bringing creation into existence, he did not abandon it to be carried away and suffer through its own nature, lest it run the risk of returning to nothing. But being good, he governs and establishes the whole world through his Word who is himself God, in order that creation, illuminated by the leadership, providence and ordering of the Word, may be able to remain firm, since it *shares in* the Word who is truly from the Father and is aided by him to exist, and lest it suffer what would happen, I mean a relapse into nonexistence, if it were not protected by the Word.[10]

The words I have emphasized translate the Greek word *metalambánousa* which can also be translated as "partakes of" or "participates in" the Word. According to Athanasius, it is partaking of the Word that enables each creature to exist and the whole creation to remain firm. From the creaturely side, then, creation is a relationship of participation, whereby all entities partake of the Word who is truly from the Source of All. They participate in the Word not only in original creation but also in what is today often called continuing creation, which Athanasius describes in terms of the Word of God "governing," "establishing," "leading," "providing," and "ordering"

---

[9] Alvyn Pettersen, *Athanasius* (London: Geoffrey Chapman, 1995), 25.
[10] *C. gent.* 41 (Thomson, *Athanasius*, 15).

the creation. Participation in the Word enables each creature to exist and the whole creation to remain firm and to flourish in one community of creation. As Athanasius puts it, the Word is "present in all things" and "gives life and protection to everything, everywhere, to each individually and to all together."[11]

In Athanasius's view, the interrelationship and cooperation of the diverse entities of the universe is the result of the work of divine Wisdom. He speaks of the Wisdom of God as bringing all the diverse creatures into balance and beautiful harmony, as keeping the oceans within their boundaries, and as providing the wonderful variety of green plants of Earth. As a musician tunes a lyre and skillfully produces a single melody from many diverse notes, so "the Wisdom of God, holding the universe like a lyre," draws together the variety of created things, "thus producing in beauty and harmony a single world and a single order within it."[12]

Through the presence of the Wisdom/Word, the various elements that make up our world are brought together in cooperation, in a kind of kinship of creation:

> Through him and his power fire does not fight with cold, nor the moist with the dry, but things which of themselves are opposites come together like friends and kin, animating the visible world, and becoming the principles of existence of bodies. By obedience to the Word of God things on earth receive life and things in heaven subsist. Through him all the sea and the great ocean limit their movements to their proper boundaries, and all the dry land is covered with all kinds of different plants, as I said above. And so that I do not have to prolong my discourse by naming each visible thing, there is nothing existing or created which did not come into being and subsist in him and through him, as the theologian says: "In the beginning was the Word and the Word was with God and the Word was God. All things were made by him and without him nothing was made" (John 1:1-3).[13]

[11] Ibid.

[12] *C. gent.* 42 (Thomson, *Athanasius,* 117).

[13] Ibid.

This idea of cooperation and kinship between creatures which is clearly developed within an ancient scientific worldview can take on new meaning today. Insights from biology have brought a new awareness of the fundamental role that cooperation plays in evolutionary history. And the ecological crisis suggests that a commitment to cooperation and a sense of kinship with other creatures will be required if human beings are to respond to the crisis of our planet and to protect its diversity of life.

Creation: In the Spirit

What is the role of the Holy Spirit in this ongoing act of creation? After neglecting the Holy Spirit in his early work, Athanasius gives expression to his Spirit theology in the *Orations against the Arians* and focuses directly on the Spirit in his *Letters to Serapion,* the first real theology of the Holy Spirit beyond the New Testament. As he had earlier insisted that the Word is proper to and belongs to (*idios*) the Father, meaning that the Word is intrinsic to the very being of the Father, so in his later work he shows that the Holy Spirit is proper to and belongs to both the Father and the Word by nature. In the process he discusses the role of the Holy Spirit in creation and articulates a comprehensive theology of creation as participation in the Trinity through the Word and in the Spirit.

Athanasius sees the Holy Spirit as indwelling in creatures as "the one who binds creation to the Word."[14] The Spirit is the bond of communion, the divine connectivity, uniting creatures to the Word and, through the Word, to the Father. In the relationship of continuous creation, the Holy Spirit enables each creature to be open to, and to receive, the creative Word. The indwelling Spirit is the enabling, empowering presence of God in creatures, who activates (*energoun*) everything that is worked by the Source of All through the Son: "For there is nothing that is not brought into being and actuated through the Word, in the Spirit."[15] In light of this theology of the Spirit, the relationship of continuous creation can be seen as

---

[14] *Ep. Ser.* 1.25 (Anatolios, *Athanasius,* 225).
[15] *Ep. Ser.* 1.31 (Anatolios, *Athanasius,* 230).

a fully trinitarian act by which God enables a world of creatures to partake of the Word in the Spirit. It is only through this participation that individual creatures exist and interact in the community of creation. In the context of his defense of the full divinity of the Spirit, Athanasius makes it abundantly clear that he sees creation as the action of the whole indivisible Trinity. It is one undivided act, but one in which each divine person acts in a way that is proper and distinctive:

> The Trinity is holy and perfect, confessed as God in Father, Son and Holy Spirit, having nothing foreign or extrinsic mingled with it, nor compounded of creator and created, but is wholly Creator and Maker. It is identical with itself and indivisible in nature, and its activity (*energeia*) is one. For the Father does all things through the Word and in the Holy Spirit. Thus the oneness of the Holy Trinity is preserved and thus is the one God "who is over all and through all and in all" (Eph 4:6) preached in the Church—"over all," as Father, who is beginning (*archē*) and fountain; "through all," through the Son; and "in all" in the Holy Spirit.[16]

Athanasius defends and proclaims both the unity and the distinction of the Trinity by appealing to the biblical texts that speak of mutual indwelling: "I am in the Father and the Father is in me" (John 14:10).[17] He sees this being in one another, which is later called *perichoresis*, as occurring in the divine life at the most profound level, because the whole being of the Word belongs to the Source of All, as radiance belongs to its light, and the river belongs to its spring. Yet, dwelling in one another, and possessing the same nature, they are truly distinct, so that the Father is Father only and not also Son. According to Athanasius's *Letters to Serapion*, the Holy Spirit possesses fully this same divine nature, and can thus enable creatures to participate in this trinitarian life through the relationships of creation and grace.

[16] *Ep. Ser.* 1.28 (Anatolios, *Athanasius*, 227).
[17] See Peter J. Leithart, *Athanasius* (Grand Rapids, MI: Baker Academic, 2011), 80–81.

Both creation and new creation, then, occur through this structure of partaking of the Word in the Spirit. God "creates and renews all things" through the Word and in the Holy Spirit.[18] In my view, Athanasius's theology of creation as participation in the life of the Trinity offers an appropriate basis for developing a contemporary theology of creation, a fully trinitarian ecological theology. This is true, above all, when this theology of creation is held together with a theology of salvation understood as a deification in which, along with human beings, the whole creation participates.

The Fruitfulness of Creation Springs from the Generation of the Word

A second insight from Athanasius that can have new meaning in today's context is his strongly held view that the generativity of trinitarian life is the ground and the source of all the fruitfulness of creaturely life. His understanding of the dynamic nature of divine life becomes apparent in the way he makes brilliant use of the various biblical names or symbols for Christ. As well as those I have already mentioned—Word, Wisdom, Image, Radiance, and Son—he also refers to River, Power, Life, and following Irenaeus, the Hand of God.[19] He interprets the fact that the Bible gives names such as Word, Wisdom, and Light to both God and Jesus Christ as pointing to their shared divine nature.

Against opponents who hold that the Son is created, Athanasius insists that Father and Son are strictly correlational. There cannot be one without the other. Just as the Son cannot be Son without the Father, so equally the Father cannot be Father without the Son. In the same way, Athanasius's other favorite images for trinitarian life reveal the Trinity's correlational nature: Light cannot ever be without its Radiance; the Spring can never be without its River; God cannot ever be without God's Word; God cannot ever be without God's Wisdom. Athanasius calls these images *paradeigmata* (symbols) and

---

[18] *Ep. Ser.* 1.24 (Anatolios, *Athanasius*, 224).

[19] Anatolios, *Athanasius: The Coherence of His Thought*, 98–100; Leithart, *Athanasius*, 41–50.

sees them as giving some revealed insight into divine being: "Since human nature is not capable of comprehension of God, Scripture has placed before us such symbols (*paradeigmata*) and such images (*eikonas*), so that we may understand from them however slightly and obscurely, as much as is accessible to us."[20]

According to Athanasius, these biblical names point to the generativity that is at the heart of trinitarian life. I have noted that he very often draws on the symbol of Christ as the Radiance (*apaugasma*) of God from Hebrews 1:3: "He is the radiance of God's glory and the exact imprint of God's very being." So, in the divine life, Athanasius sees the Light of God as shining forth eternally in the Radiance of the eternal divine Word. You cannot have Light without its Radiance, or Radiance without its Light. And the Radiance of the Light is experienced only in the Illumination of the Spirit. The Three are dynamically correlated in the one divine nature.

The different symbols qualify and correct one another. If left unqualified, the Father-Son image, for example, might be thought to imply the physicality, and the beginning in time, of human birthing. And, as recent theological analysis has shown, it leads, in the life of the church and beyond, to the false and damaging assumption that the divine persons are literally male.[21] When joined to the symbol of Light and its Radiance, it becomes clear that what is being described is an eternal dynamic correlationship between the Source of All and the eternal Word: the Father must always have the Son, as Light always has its Radiance. As Peter Leithart puts it: "The image of light and radiance thus assists in the apophatic purgation of our thoughts about God as Father and Son. One paradigm cleanses another."[22] In the divine life, the Radiance always shines from the Light, the Stream always flows from the Fountain.

This divine dynamism eternally at work in the divine life is the basis for all the fecundity of creation. Athanasius sees this as a

---

[20] Athanasius of Alexandria, *Orations against the Arians* (= *C. Ar.*) 2.32, in Anatolios, *Athanasius*, 127.

[21] See Elizabeth Johnson, *She Who Is: The Mystery of God in Feminist Theological Discourse* (New York: Crossroad, 1992), 33–41.

[22] Leithart, *Athanasius*, 46.

powerful argument against those who deny the eternity of the Word. He points out that the wonderful fruitfulness of God's creation that is evident all around us must point back to a generativity in divine life that is eternal. Creation comes to be as a free act of God in time, but it can only be grounded in the eternal possibility of creating in the triune God. The fecundity of creation can spring only from the eternal dynamic fecundity of divine life.

But if, as his "Arian" opponents supposed, the creative Word/ Wisdom of God is a creature who has a beginning, then this would undermine what Athanasius calls the eternal "generative nature" of God.[23] Then the divine essence would be understood as barren rather than fruitful, as a light without its radiance, as a fountain without its flow of living water. To say "there was when the Word was not" is to say that there was when the divine Fountain was dry and destitute of Life and Wisdom. Athanasius points to what he sees as the emptiness at the heart of his opponents' position:

> In accord with them, let not God be of a generative nature, so that there may be no Word nor Wisdom nor any Image at all of his own essence. For if he is not Son, then neither is he Image. But if there is no Son, how then do you say that God is Creator, if indeed it is through the Word and in Wisdom that everything that is made comes to be and without which nothing comes to be, and yet, according to you, God does not possess that in which and through which he makes all things (cf. Wis 9:2; Jn 1:3; Ps 104:20, 24). But if, according to them, the divine essence itself is not fruitful but barren, like a light that does not shine and a fountain that is dry, how are they not ashamed to say that God has creative energy?[24]

God is a Light with its everlasting Radiance that enlightens us in the Spirit. God is a Spring ever pouring forth a River of living water from which we creatures drink in the Spirit. God always has God's own divine Wisdom, of whom we partake in the Spirit. God is a Father eternally begetting the Son in whom we participate by

---

[23] *C. Ar.* 2.2 (Anatolios, *Athanasius*, 111).
[24] Ibid.

adoption as God's children in the Spirit.[25] Such, Athanasius argues, is the "correlation (*sustoichia*) and the unity of the Holy Trinity."[26] Those who deny the full and eternal divinity of the Word or the Spirit deny the dynamic life of God, the divine generativity, that is the very ground of the creation and salvation of a world of creatures.

In the first section of this book I mentioned biblical texts where God is imaged as Mother. These texts point to God's faithful, tender love—"As a mother comforts her child, so will I comfort you" (Isa 66:13) and invite human beings to radical trust in God—"But I have calmed and quieted my soul, like a weaned child with its mother" (Ps 131:2). Can this image of God as Mother function in a fully trinitarian theology? If Athanasius uses the images of Light-Radiance and Spring-Stream as "apophatic qualifiers" of the Father-Son image, can we today make similar use of the image of mothering in a fully trinitarian theology? For this to be true to the kind of theology Athanasius promotes, I think it would need to meet two conditions: it would need to be capable of expressing divine generativity on the one hand and the correlationality of the divine persons on the other. In other words, it would need to express the relations of origin. Clearly, the image of mothering is as fully generative as fathering—perhaps more so. The contributions of carrying a child in the womb and giving birth are more obviously generative than the contribution of the father. And the Mother-Child, or Mother-Wisdom/Word, relation can be easily seen to be fully correlational. This means that the Source of All in the Trinity can be imaged, prayed to and loved, not only as Abba/Father, but also as Mother.

The fecundity of the creation, the bringing into existence of a world of stars and galaxies, of seas, land, and atmosphere, of microbes, insects, animals, plants, and human beings, is grounded in the divine dynamic fecundity that is intrinsic to the very being of God. As Athanasius sees it, the free creation of a world of creatures is grounded in the eternal generation of the Word. Of course, it can

---

[25] *Ep. Ser.* 1.19 (Anatolios, *Athanasius*, 217–18). See also Athanasius, *On the Council of Nicaea* (= *Decr.*), 23–24 (Anatolios, *Athanasius*, 200–201).

[26] *Ep. Ser.* 1.20 (Anatolios, *Athanasius*, 219).

be added that it also springs from the eternal proceeding of the Holy Spirit, the procession that Athanasius calls the "shining forth" of the Spirit.[27] God's eternal, dynamic, endlessly relational life involves both. The generativity of the divine nature finds free expression in all the beauty, diversity, and strangeness of the universe of creatures.

Athanasius's thought, developed in response to the Arian challenge, can offer new and rich meaning in an ecological age. The trinitarian God that he defends is a God of boundless fruitfulness. God is fruitful by nature. The fruitfulness of the natural world, of the rain forests of Earth with all their interdependent species of plants and animals, of a handful of topsoil with its millions of busy microbes, of this blue wren I see in front of me, are all grounded in the dynamic generativity and fruitfulness of God the Trinity. They spring from the Source of All, the divine Mother/Father, endlessly generating the eternal Wisdom/Word and breathing forth the Spirit of life and love.

Creation's Place within the Divine Delight

Athanasius recognizes the limited nature of human speech about God and so, for him, language about things being *in* God, or God *containing* creation, is analogical. God is not literally in space at all. But, granted this, Athanasius refers to the words attributed to Paul in Acts, "In him we live and move and have our being" (17:28), and applies them to the Word. Thus, he says, all creatures exist in the Word, through whom the universe has "light, life and being."[28]

Athanasius speaks of the Word and Savior "through whom the Father orders the universe and contains and provides for all things."[29] The image of God containing the universe through the Word appears again when Athanasius discusses the difference between God's creative action and human creativity: "Moreover, humans, who are of themselves incapable of being, exist in place and

---

[27] Ibid., 220.

[28] Athanasius of Alexandria, *On the Incarnation* (= *Inc.*) 41 in Thomson, *Athanasius*, 237.

[29] *C. gent.* 47 (Thomson, *Athanasius*, 131).

are encompassed and sustained in the Word of God, while God is self-existent, containing all things and being contained by none, and is in all things according to his own goodness and power, but outside all things according to his own nature."[30]

While God the Trinity can be imaged as "containing" the universe of creatures, God is not contained by anything at all. It is typical of Athanasius's language to say that God is "outside" of all things according to the divine nature, in order to express his conviction that the divine nature is beyond all created being and all comprehension. This, of course, does not mean that God is absent from creatures. As Athanasius also insists, God is not divided. When God is present in love, empowering the existence and life of all things, God the Trinity is fully and immediately present. The divine essence, however, remains beyond creatures.

Athanasius has a related way of pointing to the "place" of creation in God. He thinks of creation as occurring within the interpersonal love and delight of the divine persons. In Luke we find Jesus presented as rejoicing in the Spirit, and celebrating and testifying to the mutual knowledge of the Father and the Son (Luke 10:21-22), a theme also found in John (John 10:15). Reflecting on this relationship, Athanasius recalls the words of divine Wisdom in the book of Proverbs: "Then I was beside him, like a master worker; and I was daily his delight" (Prov 8:30). He sees this biblical teaching of God's delight in Wisdom as pointing to the eternal delight within the life of the Trinity, and he locates God's joy in creation within this mutual divine delight.

His argument is that God's delight in Wisdom can be an eternal gladness only if divine Wisdom is eternally with the Father. The Arian view that Wisdom had a beginning would deny this possibility of mutual rejoicing in God. In Athanasius's view, the eternal delight of the life of the Trinity is the foundation for God's joy in creaturely reality:

> When was it then that the Father did not rejoice? But if he has always rejoiced, then there was always the one in whom he

---

[30] *Decr.* 11 (Anatolios, *Athanasius*, 188).

rejoiced. In whom, then, does the Father rejoice (cf. Prov 8:30), except by seeing himself in his own image (*eikoni*), which is his Word? Even though, as it has been written in these same Proverbs, he also "delighted in the sons of people, having consummated the world" (Prov 8:31), yet this also has the same meaning. For he did not delight in this way by acquiring delight as an addition to himself, but it was upon seeing the works that were made according to his own image, so that the basis of this delight also is God's own Image.[31]

Athanasius's central point is that in spite of his opponents' misuse of Proverbs 8:22, "He created me as the beginning of his ways," God's delight in Wisdom does not have a beginning but is an eternal rejoicing. The mutual delight of Father and Word in the Spirit is intrinsic to the divine being, and the biblical theme of God's delight in creatures is situated within this mutual gladness. God's joy in human beings and other creatures, then, is not an addition to the divine being, but is "an inclusion of the creation into the eternal mutual delight of the being of the Father and the Son."[32] God's relationship to creation is embraced within the divine joy of the Trinity.

The creation of a world of creatures from nothing is grounded in this fecundity of inner trinitarian life. Creation takes place within the mutual love and bliss of the divine persons. As Anatolios puts it, God's delight in creatures is "an inclusion of creation within the eternal mutual delight of the being of Father and Son."[33] The Holy Spirit enables the mutual delight of the divine persons to be sharable and brings about creation as the site of the extension of the Father-Son relation beyond the divine being. Anatolios goes on to say "Such a trinitarian account of creation speaks to our contemporary ecological crisis, leading us to see that a destructive posture towards creation is blasphemous in its dishonouring of the Father-Son delight and the Spirit's gift-giving of that delight."[34]

---

[31] *C. Ar.* 2.82 (Anatolios, *Athanasius*, 175).

[32] Khaled Anatolios, *Retrieving Nicaea: The Development and Meaning of Trinitarian Faith* (Grand Rapids, MI: Baker Academic, 2011), 153.

[33] Ibid., 118, 153–54, 288.

[34] Ibid., 288.

The Immediacy of the Triune God to Each Creature

If God is immediately and lovingly present to each creature, then this suggests an attitude of deep respect for other species and for individual creatures. It would have consequences for ecological ethics and action. In this final section I will point to Athanasius's strong argument for the immediacy of the triune God to creatures. His reasons, of course, are not ecological. His concern, as always, is for the full divinity of the Word and Spirit.

Athanasius opposes the assumption, widespread in Platonic philosophy and shared by his opponents, of the necessity for some kind of created intermediary between the world of creatures and the all-holy God. Athanasius completely rejects the notion of this kind of intermediary. While his model of participation is broadly Platonic, it is developed in a distinctively Christian way. He certainly insists, as I have pointed out, on both the radical otherness of the Creator and on the intrinsic poverty of being of all creatures. He thus sees an infinite difference between finite creatures and the Creator. How, then, is this gulf to be bridged?

In standard Platonic views, the answer is through secondary intermediate figures, such as the Demiurge and the world of Ideas, or the Logos, or the Soul. Creatures participate in the intermediary, while the intermediary participates in God but is not God. For Athanasius's Christian opponents, the divine dignity and holy otherness of God are such that there can be no direct relation between God and creatures. Any direct connection would diminish or demean God. And finite creatures would not be able to endure the blazing touch of the infinitely holy God. So, the opponents reason, God *creates* the Word as a mediator to carry out God's purposes in creation and salvation. In their view, as it is summarized by Peter Leithart, the Word of God "serves as a buffer between God and creation."[35]

For Athanasius there is no such buffer. This sharply distinguishes his thought from various forms of Platonic philosophy and from Christian thinkers like Arius, Eusebius, and Asterius. Athanasius agrees with them on the radical otherness of the Creator and on

---

[35] Leithart, *Athanasius*, 91.

the biblical conviction that God engages with creation through the Word. But he insists that the Word is not a creature, but possesses fully the Father's very essence and, precisely as fully divine, bridges the gap between Creator and creatures in loving condescension: "For they would not have withstood his nature, being that of the unmitigated splendour of the Father, if he had not condescended (*sunkatabas*) by the Father's love for humanity and supported, strengthened, and carried them into being."[36] It is important to grasp that in Athanasius's usage, condescension does not connote patronizing behavior. Rather, Athanasius is applying its literal meaning: the Word of God "comes down" to be with us in immediate presence out of compassionate love.

While his opponents associated transcendence with the Source of All and immanence with the created Word, Athanasius saw the Word, and also of course the Spirit, as fully divine and as completely transcendent. But, as Anatolios points out, Athanasius also radically transforms the idea of divine transcendence by means of the biblical categories of divine mercy and loving kindness. Love and mercy, then, are divine attributes by which God can transcend God's own transcendence.[37] As fully divine and possessing the divine capacity for mercy and loving kindness, the Word and the Spirit can be unthinkably close to creatures in the relationship of creation. Because of divine mercy and love, no creaturely mediation is needed. And the character of God in creating thus accords with the character of God revealed in the incarnation.[38]

The radical ontological distinction between God and all creatures is not bridged by a created intermediary, but solely from God's side, in a loving generosity that is itself divine. Because Word and Spirit are one with the Father's essence, the Word's mediation in the Spirit means, of course, that the Father, the Source of All, is also immediately present to each creature.[39] As Athanasius puts it,

---

[36] *C. Ar.* 2.64 (Anatolios, *Athanasius*, 157–58).

[37] Anatolios, *Retrieving Nicaea*, 104.

[38] Ibid.

[39] Anatolios, *Athanasius: The Coherence of His Thought*, 113.

using again a favorite image, the one who experiences the Radiance is enlightened by the Sun itself and not by any intermediary.[40]

In Athanasius, then, we find a theology of the immediate presence of the triune God to all creatures. Created entities participate immediately in trinitarian life. They do not possess the divine nature, but participate only by the gracious act of divine generosity and love by which God bridges the ontological gap between the infinitely other God and a world of finite creatures. The bridging of the abyss between creation and Creator cannot be the work of a creature, but only that of the Source of All, through the Word and in the Spirit.[41] The Word can be thought of as a mediator, but only as a fully divine mediator of fully trinitarian presence. Only God can relate the world of creatures to God's self.

What Athanasius clearly brings to light is that the true nature of the God-creature relationship—above all its radical immediacy—can be understood only when Word and Spirit are understood as fully divine. A theology of God as Trinity enables us to glimpse the immediacy of the relationship between God and all creatures. Things have their existence only from this relationship, which means that ultimately "Athanasius's perspective is that of a relational ontology."[42] Every creature on Earth, every whale, every sparrow, and every earthworm exists by participation in the Mother/Father through the Word and in the Spirit—"not one of them is forgotten in God's sight" (Luke 12:6).

In this chapter I have been proposing that in boundless generosity the Source of All continuously creates each creature through the Word and in the Spirit, that the fruitfulness of this creative act springs from the generativity of the divine life, that the whole universe of creatures exists with the relations of mutual delight of

---

[40] *C. Ar.* 3.14, in William Bright, *The Orations of Athanasius against the Arians* (Oxford: Clarendon Press, 1884), 169.

[41] See Anatolios, *Athanasius: The Coherence of His Thought*, 162: "Athanasius's whole logic was averse to the notion of a created mediation between God and creation, since it is exclusively a divine characteristic to be able to bridge the distance between God and creation."

[42] Ibid., 208.

the Trinity, and that in continuous creation the divine persons are immediately present to each creature, enabling each to exist and act within the community of creation. The God whose transcendence involves self-humbling love is immanent to a world of creatures in a way that no creature could ever be. This line of thought suggests that all the diverse creatures of our planet are the place of trinitarian presence, the gift of the triune God, and that they have their own intrinsic value within the one community of creation.

Chapter 2

Trinity in Act

Deifying

Many people today find difficulties with the Christian theology of salvation in Christ. Part of the problem, I believe, is that theories of salvation that have been long preached and taught in the life of the Western church are no longer functioning as authentic good news. These include, in the Catholic world, the satisfaction theory of Saint Anselm of Canterbury and, in the Protestant world, the substitutionary atonement theory of the Reformation. While these theories have been meaningful to many generations of Christians, they are easily misunderstood in the contemporary world, leading to distorted and damaging views of God. In extreme cases the God of Jesus is misrepresented at the popular level in blasphemous terms as one who demands satisfaction in blood.

In suggesting that there is a need to go beyond these theories today, I am not advocating abandoning the words used in the New Testament to communicate what God does for us in Christ and the Spirit, words like salvation, sacrifice, redemption, reconciliation, justification, sanctification, new creation, freedom, transformation, new life, and adoption. Each is based on a different image, and each has its own resonance in ordinary life as well as in the Scriptures. It seems that for the New Testament authors, this rich variety was necessary because no one image or concept was sufficient to express what is finally beyond all our words: God's love poured out for us in Jesus' life, death, and resurrection and in the giving of the Spirit.

An important distinction can be made between images of salvation in Christ, such as those found in the New Testament, and overarching theological theories of salvation.[1] Images can exist alongside each other, each providing a partial insight into the Christ event without necessarily forming a comprehensive theory. A theory, or a theology, by contrast, attempts to provide a coherent account of the meaning of Jesus' life, death, and resurrection. Biblical images and words can offer rich theological meaning within different overarching theories.

In today's context, I see two very different theories as having unique potential to enable new generations to see something of the meaning of Christ for our world. One is the theology of liberation, explored in the last century by Gustavo Gutierrez and others, and the other is that of deification, articulated long ago by Athanasius, among others. In taking up Athanasius's theology of deification, I will begin by discussing his understanding of the deification of the human person. Then I will ask about what it is that Athanasius sees as new in deification, over and above the divine act of creation discussed in the last chapter. In the third section I will focus directly on the issue that is fundamental to this book, the application of the theology of deification to the rest of the natural world. This will lead to a final section that draws out some conclusions from this theology of deification.

Deification and the Transformation of the Human

Athanasius sees human beings as, at the original creation, being freely given the special grace of participating in the Word and possessing eternal life. Humans were made according to the image (*kat' eikona*) of the Word, the true and eternal Image of God. But they sinfully rejected this grace. They lost the stability of being according to the Image and forfeited the gift of eternal life, and thus faced death. In the incarnation of the Word, the true Image of God comes to repair and restore the image of God in humanity and to overcome death.[2]

---

[1] Peter Schmiechen, *Saving Power: Theories of Atonement and Forms of the Church* (Grand Rapids, MI: Eerdmans, 2005), 5.

[2] See Khaled Anatolios, *Retrieving Nicaea: The Development and Meaning of Trinitarian Doctrine* (Grand Rapids, MI: Baker Academic, 2011), 107.

It is important to note that when Athanasius speaks of the incarnation, he is not referring primarily to the beginning of Jesus' earthly life, but to the whole event of the life, death, and resurrection of the Word made flesh. The Word takes our finite, fleshly humanity, damaged by sin and subject to death, in order to transform it from within. Athanasius sees the incarnation as the response of a loving, benevolent God to the predicament caused by sin that had left the world dominated by sin and death and lacking true knowledge of God. In this context, Athanasius names two reasons for the incarnation. He sees the overcoming of death as "the primary cause of the incarnation of the Saviour."[3] The second is that we might know "the Word of God who was in the body, and through him the Father."[4]

Athanasius's reasons for the incarnation hold, I believe, even if one does not adopt his view that biological death comes about because of human sin. He sees the death of Jesus, in light of the resurrection, as bringing us forgiveness, healing, liberation from the power of evil, resurrection life, and, in a particular way, freedom from the fear of death (Heb 2:14-15). Inspired by Hebrews, and it seems by the Eucharist, he speaks of the cross of Jesus in the liturgical language of offering: He "surrendered his body to death in place of all and offered it to the Father."[5] Athanasius takes up a whole range of biblical concepts for the meaning of Christ that he finds in Paul and in Hebrews and provides the overarching vision of deification that I explore here.

In Athanasius's view of the incarnation, the compassionate condescension and radical *kenōsis* evident in Jesus' life and death are not the sign of a lack of divinity, as his opponents supposed. They are, rather, the revelation of the divine, transcendent nature of God as self-humbling love. Anatolios makes this point:

---

[3] Athanasius of Alexandria, *On the Incarnation* (= *Inc.*), 10 in *Athanasius: Contra Gentes and* De Incarnatione, trans. Robert Thomson (Oxford, UK: Clarendon Press, 1971), 159.

[4] *Inc.* 14 (Thomson, *Athanasius*, 169).

[5] *Inc.* 8 (Thomson, *Athanasius*, 153).

These human limitations and sufferings are seen to be not reflective of a lower level of transcendence but rather disclosive of divine perfection and transcendence reconceived as loving condescension and self-abasement. This involves the insertion of the categories of *kenōsis*, compassion and self-humbling into the canon of absolute divine perfections. In this way the creaturely primacy of Christ manifest in his life of obedience and compassionate suffering comes to be seen as a reflection and not a mitigation of absolute divine primacy. Compassionate self-abasement becomes both a divine attribute and a characterization of the humanity of Christ.[6]

While others might interpret Jesus' humility and obedience as the subordination of the Word to the Source of All, for Athanasius, the self-humbling found in Jesus Christ truly reveals the proper understanding of the transcendent God. What I pointed out about the divine act of creation in the last chapter is found to be even more striking in relation to the incarnation: the concept of divine transcendence is transformed by the biblical categories of love and mercy. God is divinely, transcendently, self-humbling. It is the humility of God in Christ that brings about our deification.

For Athanasius, deification means that through the incarnation of the Word and by the grace of the Holy Spirit, we are transformed in Christ and assimilated to him in his relationship to Abba/Father. By grace we participate in the triune God. Athanasius first uses the word deification to describe this transformation in his *On the Incarnation*, where he says of the Word made flesh: "For he became human that we might become divine."[7] With this phrase he sums up the meaning of the incarnation simply and succinctly. The incarnation is for the sake the deification of creatures.

Athanasius uses deification language very often in his later, explicitly anti-Arian, writings. In these texts, his principal argument for the divinity of the Word is based on the structure of salvation understood as deification: Only if the Word is truly divine can we be truly deified by the Word. Thirty times he uses the verbal form "to

---

[6] Anatolios, *Retrieving Nicaea*, 287.

[7] *Inc.* 54 (Author's translation of the Greek text in Thomson, *Athanasius*, 153).

deify," with the Greek word *theopoieō*. He also uses "deification" as a noun, coining the Greek word *theopoiēsis*. Many who come later in the Greek tradition use an alternative word for deification, *theosis*.[8]

Athanasius deepens the theological meaning of deification by insisting, against the earlier view of his predecessor Origen (c. 184–253), and also against later "Arian" views, that the Word of God is *not* deified, but is always the source of deification for creatures: "So he was not a human being and later became God. But, being God, he later became a human being in order that we may be divinized."[9] Athanasius notes that even in times past, long before the birth of the Savior, the eternal Word of God was already the source of deification, for Moses and for such other faithful ones who were adopted and deified as children of God.[10]

In developing the theology of deification, Athanasius builds upon the work of his predecessors, particularly Irenaeus (130–202). It has been said that he takes up a theme that is more or less secondary in earlier writers and makes it the central idea of his own theology.[11] It is not only that he uses deification language more frequently than his predecessors but also that he does a great deal to clarify its meaning. One of the ways he does this, as Norman Russell has shown, is by constantly pairing the word deification with a series of other explanatory words. He uses the following words as synonyms that function as equivalents to deification and help to explain its meaning: "adoption," "renewal," "salvation," "sanctification," "grace," "transcendence," "illumination," and "vivification."[12]

---

[8] Norman Russell, *The Doctrine of Deification in the Greek Patristic Tradition* (Oxford, UK: Oxford University Press, 2004), 167–68. Mainly in his early work, Athanasius also uses the word deification about twenty times in a negative sense to describe the way pagans make creatures into their gods.

[9] Athanasius of Alexandria, *Orations against the Arians* (= *C. Ar.*) 1.39 in Anatolios, *Athanasius* (New York: Routledge, 2004), 96.

[10] Ibid.

[11] Jeffrey Finch, "Athanasius on the Deifying Work of the Redeemer," in *Theōsis: Deification in Christian Theology*, ed. Stephen Finlan and Vladimire Kharlamov (Eugene, OR: Pickwick Publications, 2006), 104.

[12] Russell, *The Doctrine of Deification*, 177–78.

In Athanasius's thought, deification expresses the teaching of the Second Letter of Peter that in Christ we become "participants of the divine nature" (2 Pet 1:4). Of course, we never possess the divine nature, but we partake of it by grace. We participate in Christ by the grace of the Spirit, and find our human fulfillment in the life of the triune God.

But the very frequent linkages Athanasius makes with adoption show that he identifies deification particularly with Paul's theology of adoption in Christ: we are God's adopted daughters and sons, and the Spirit dwells in us so that we pray "Abba! Father!" (Gal 4:4-7; Rom 8:14-16). This, of course, is closely connected to the related theme in the Gospel of John: we become "children of God" (John 1:12), we are "born . . . of God" (John 1:13) or are "born of the Spirit" (John 3:8). In Athanasius's theology, the concept of deification is linked with the New Testament idea of adoption more clearly, more frequently, and more directly than his predecessors.[13]

When Origen discusses deification, he speaks of the deification of the mind (*nous*). With Athanasius, by contrast, it is more the flesh, the body, humanity, or simply creation itself that is spoken of as transformed by the Word made flesh. This fleshly emphasis can have helpful implications in the present context, suggesting that deification in Christ can be seen as involving not only the human mind and spirit but also the transfiguration of the whole material, fleshly creation.

Because the Word of God is eternally divine, Athanasius completely rejects the view that this Word needs to be deified. He nevertheless sees a vitally important deification at work in Jesus Christ, the Word made flesh. In his view, the created bodily humanity of Jesus *is* deified by its union with the Word. It is precisely this deification of Christ's creaturely humanity by the Word, in the power of the Holy Spirit, that enables the deification of our humanity:

---

[13] Vladimir Kharlamov, "Rhetorical Applications of *Theosis* in Greek Patristic Theology," in *Partakers of the Divine Nature: The History and Development of Deification in the Christian Traditions*, ed. Michael J. Christensen and Jeffery A. Wittung (Madison, NJ: Fairleigh Dickenson University Press, 2007), 121.

For the Word was not lessened by his taking a body, so that he would seek to receive grace, but rather he divinized what he put on, and, what's more, he gave this to the human race.[14]

In Jesus, the Word of God takes finite human flesh as the Word's very own, and as a result, we creatures of flesh participate in deification with him. For Athanasius, there is a solidarity in the flesh by which both the body of Christ and our bodily humanity are deified:

> He took to himself the body that was human and had a beginning so that he who is its creator may renew it and thus divinize it in himself and lead all of us into the kingdom of heaven, in accordance with his own likeness . . . . But humanity would not have been deified if joined to a creature, or unless the Son was true God. And humanity would not come into the presence of the Father unless the one who put on the body was his true Word by nature. Just as we would not have been freed from sin and the curse unless the flesh which the Word put on was human by nature—for there would be no communion for us with what is other than human—so also humanity would not have been deified unless the Word who became flesh was by nature from the Father and true and proper (*idios*) to him. Therefore the conjoining that came about was such as to join what is human by nature to what is of the nature of divinity, so that humanity's salvation and deification might be secured.[15]

It is central to Athanasius's thought that the Word is proper (*idios*) to the being of the Father, sharing the divine essence. And in the incarnation, this Word, who is eternally the Father's very own, now "appropriates" created human nature, making it the Word's very own. Henceforth humanity and creatureliness are not external to the Word but have become the Word's very own.

Some contemporaries question the theology of deification on the basis that it might obscure what it is to be truly human. Athanasius's response to this, I believe, would be to say that to be fully human, to be deeply true to our humanity, is to participate in God.

---

[14] *C. Ar.* 1.42 (Anatolios, *Athanasius*, 99).

[15] *C. Ar.* 2:70 (Anatolios, *Athanasius*, 163).

His view of the human would be similar to that summed up in the famous saying of Augustine at the beginning of his *Confessions*: "You have made us for yourself, O Lord, and our hearts are restless until they rest in you."[16] The deification of humanity is not about changing human nature into something other than it is, but about becoming fully human in a way that is faithful to God's intention. As Andrew Louth says, it is not about going beyond the human, but about becoming "the human partners of God" and the participants in divine life that we were created to be. Deification, then, is not a transcending of the human, but the true fulfillment of the human.[17] It is the purpose for which human beings were made.

For Athanasius, deification is a radical ontological transformation in creaturely reality. He also discusses what is sometimes called the ethical aspect of deification, our growth in holiness. But his theological emphasis is on the divine transformation already at work in humanity and in the world. Through the flesh assumed by the Word, God communicates divine life to all flesh in principle. This divine life is transmitted in practice to individual human beings through the gift of the Spirit in the life of grace. Something is already given in Christ, new creation has begun, but our participation involves a free human response. For the Christian community, as Athanasius sees it, this deification is embraced and lived out in sacramental and communal life, in prayer and asceticism, and in witness to Christ.

The Newness of Deification

If deification is a participation in trinitarian life, a partaking of the Word in the Spirit, how is this new relationship distinct from the relationship of creation which is already a participation in the life of the Trinity? What is new in this deification that comes about through the incarnation? A first fundamental answer is that what is radically new is the resurrection of the crucified Christ as the source of new creation for human beings and the rest of creation and the

[16] Augustine, *Confessions* 1.1.

[17] Andrew Louth, "The Place of *Theosis* in Orthodox Theology," in *Partakers of the Divine Nature*, 39.

outpouring of the Pentecostal Spirit. In the Spirit we participate in the glorious newness of the crucified and risen Christ.

For Athanasius there is something new about the *interior* nature of God's self-giving to creatures in the incarnation of the Word. Anatolios points to three ways in which Athanasius highlights this new interiority.[18] The first is his emphasis on the idea that deification provides a new *security* in our participation in divine life. What is given to us in Christ is not simply a return to the instability that characterized the relationship of creation. Rather, through Christ's saving work, we now remain securely in God. Athanasius takes up the language of "remaining" or "abiding" (*menein*) in Christ, widely used in the Gospel of John.[19] Because of the incarnation, we are united to God through the Word who is God, and so we can be sure that we can "remain" securely in the divine communion in stability and peace. We have reason to hope and trust that, by the grace of God, we will remain in Christ forever through our participation in his resurrection from the dead.[20]

A second aspect of this new interiority is found in Athanasius's idea that the grace given in Christ is newly "internal" to us as creatures. He often makes the point that the divine essence is "outside" or beyond the world of creatures—meaning that we never possess the divine essence. But when God seeks to save creatures, it is not from outside, by an external divine decree. It is true that in the beginning God brought creation into existence by a command of the will, "by a nod only." But once creatures existed and sinful humanity needed to be cured and liberated, God acted to bring salvation *internally*. The Savior came in the body to transform bodily existence and death itself from within.[21] What this means is that our communion with God has now become something internal to us in our creatureliness and our humanity, since we have now become "Worded through the Word of God, who became flesh for our sake."[22]

---

[18] Anatolios, *Athanasius*, 61–70.

[19] As one of many examples, see John 15:3-10.

[20] *C. Ar.* 2.69 (Anatolios, *Athanasius*, 162–63).

[21] *Inc.* 44 (Thomson, *Athanasius*, 245–47).

[22] *C. Ar.* 3.33 (Anatolios, *Athanasius*, 66).

A third way of expressing the new interiority of what happens in Christ is through the idea that the Word now *"belongs to"* or is *"proper to"* us as creatures. These English expressions attempt to translate Athanasius's key word *idios*. He often uses this word to insist against his opponents that the Word of God is not "outside" the Father's but "proper" to the Father. The meaning is that the Son belongs absolutely to the Father—Father and Son are eternally correlative. What happens in the incarnation is that this Word now makes fleshly human reality the Word's very own. The Word appropriates (*idiopoieō*) creaturely existence, making it proper to the Word's self. Anatolios writes: "The startling *novum* of the Incarnation is that the eternal divine Word whose nature is 'outside' the created order 'appropriated' created human nature so that henceforth humanity was not external to his being but has become his very own."[23] By this appropriation we are no longer bound by sin and death, but are newly transformed in resurrection life.

At the heart of this new interiority is the Holy Spirit. Deification itself is simply the transforming gift of the ever-new Spirit of God. As Athanasius sees it, in the incarnation, Jesus in his humanity becomes the receiver of the Holy Spirit and is deified by the Spirit. Through Jesus and his openness to the Spirit, we too are enabled to become coreceivers of the Spirit. Born again by the grace of the Spirit, we are enfolded in the inner life of the Trinity, taken up in the position of the Word in relation to the Father.[24] As the personal human beings we are, as fully interpersonal, and as interrelated with the wider community of creation, we are made into the image of the Word, taken into the divine life of the Trinity, so that our creaturely humanity flourishes in its true home.

The Transfiguration of the Rest of the Natural World

Orthodox theologian Andrew Louth, writing of the Eastern tradition of deification, suggests that we might see this deification as the fulfillment of the divine act of creation, rather than simply as con-

---

[23] Russell, *The Doctrine of Deification*, 67.

[24] *C. Ar.* 1.46 (Anatolios, *Athanasius*, 103–4); *C. Ar.* 2.59 (Anatolios, *Athanasius*, 152–53). See Anatolios, *Retrieving Nicaea*, 125.

nected to the rectification of human sin.[25] Deification, then, is not just the equivalent of the redemption of sinful human beings, but belongs to a broader conception of the divine economy. Louth explains this broader conception by means of an image of two arches. The greater arch stretches from creation to deification. The lesser arch leads from human sin to redemption. The greater arch represents "what is and remains God's intention: the creation of the cosmos that, through humankind, is destined to share in the divine life, to be deified."[26] Because of human sin, humanity has failed God in its participation in the work of the greater arch—hence the need for redemption.

What of Athanasius? How does he see the relationship of deification to the rest of the natural world? As I have already said, Athanasius's concern is not that of a twenty-first-century ecological theology. His constant preoccupation is the defense of the full divinity of the Word and the Spirit. When he discusses deification, his primary focus is often on the human. After all, his basic argument is *ad hominem*, directed at his human readers, pointing out that the reality of their deification depends upon the true divinity of the Word and Spirit who brings it about.

But like the biblical tradition before him, Athanasius sees human beings not as isolated individuals, but as in relationship with each other, and as interrelated with the rest of the natural world. And, like the biblical texts to which he constantly refers, such as Romans 8:19-23 and Colossians 1:15-20, Athanasius clearly sees this wider natural world as participating with humanity in liberation and transformation in Christ. He writes, for example, of the cosmic meaning of the incarnation in these words:

> The Word, in the Spirit, fashioned and joined a body to himself, wishing to unite creation to the Father and to offer it to the Father through himself and to reconcile all things in his body, "making peace among the things of heaven and the things of earth" (cf. Col 1:20).[27]

---

[25] Andrew Louth, "The Place of *Theosis* in Orthodox Theology," 34–35.
[26] Ibid.
[27] *Ep. Ser.* 1.31 (Anatolios, *Athanasius*, 232).

In the last chapter, I pointed to Athanasius's rich and broad theology of creation, whereby all creatures partake of the Word in the Spirit. When he considers the incarnation, he consistently sees it as an event between God and the whole creation.[28] So even when he speaks quite specifically of the adoption of human beings, he often brings the wider creation into the discussion. In his *Letters to Serapion*, for example, he writes of the human beings anointed with the Holy Spirit: "When we are sealed in this way, we properly become sharers in the divine nature, as Peter says (2 Pet 1:4), and so the whole creation *participates of* (*metechei*) the Word, in the Spirit."[29] Long ago, C. R. B. Shapland, in his translation of these letters, commented that Athanasius's reference to the whole creation sharing in *salvation,* by partaking of the Word in the Spirit, seems a natural extension of his statement in *Against the Nations* (quoted in the last chapter), that creatures owe their very *existence* to their partaking of the Word.[30]

Athanasius shows no interest in separating humans from the rest of nature, or in making sharp distinctions between human beings and other creatures in their participation in deification. He brings in the wider creation as a matter of course, assuming that the whole creation will share with human beings, in its own specific ways, in glorification and deification. In another example from the *Letters to Serapion*, to which I referred briefly in the last chapter, Athanasius is making the argument that the Holy Spirit, in whom the Word adopts and deifies creatures, cannot be a creature:

> Therefore, it is in the Spirit that the Word glorifies creation and presents it to the Father by divinizing it and granting it adoption. But the one who binds creation to the Word could not be among the creatures and the one who bestows sonship upon creation could not be foreign to the Son. Otherwise, it would be necessary to look for another spirit to unite this one to the Word. But that

[28] Khaled Anatolios, *Athanasius: The Coherence of His Thought* (New York: Routledge, 1998).

[29] *Ep. Ser.* 1.23 (Anatolios, *Athanasius,* 223).

[30] C. R. B. Shapland, *The Letters of Saint Athanasius Concerning the Holy Spirit* (London: Epworth Press, 1951), 124–25, n. 15.

is senseless. Therefore, the Spirit is not among the things that have come into being but belongs (*idion*) to the divinity of the Father, and is the one in whom the Word divinizes the things that have come into being. But the one in whom creation is divinized cannot be extrinsic to the divinity of the Father.[31]

For Athanasius, the Holy Spirit is the one who binds all creatures to the Word, the one in whom human beings are made daughters and sons, and the one in whom the Word deifies creation. Alongside such references to the deification of creation, there are times when Athanasius is fully explicit about the wider creation participating in Christ. An example is found in his second *Oration against the Arians*, where he refers to Romans 8:19-23 and Colossians 1:15-20, and unambiguously sees the rest of the natural world as sharing in the liberation that comes in Christ's resurrection. Defending the divinity of Christ against his opponents, he writes:

> The truth that refutes them is that he is called "firstborn among many brothers" (Rom 8:29) because of the kinship of the flesh, and "firstborn from the dead" (Col 1:18) because the resurrection of the dead comes from him and after him, and "firstborn of all creation" (Col 1:15) because of the Father's love for humanity, on account of which he not only gave consistence to all things in his own Word but brought it about that the creation itself, of which the apostle says that it "awaits the revelation of the children of God," will at a certain point be delivered "from the bondage of corruption into the glorious freedom of the children of God" (Rom 8:19-21). The Lord will be the firstborn of this creation which is delivered and of all those who are made children, so that by his being called "first," that which is after him may abide, united to the Word as to a foundational origin and beginning.[32]

Throughout this text, Athanasius links creation's deliverance to the participation of human beings in resurrection life. Then in the last sentence, he brings them together again in a neat summary. The

---

[31] *Ep. Ser.* 1.25 (Anatolios, *Athanasius*, 225).
[32] *C. Ar.* 2.63 (Anatolios, *Athanasius*, 157).

risen Christ is the firstborn, the "foundational origin and beginning," of both creation's deliverance and of human beings being made God's children.

In his *Letter to Adelphius*, written around 370, Athanasius speaks of Christ as "the Liberator of all flesh and of all creation (cf. Rom 8:21)." Toward the end of the same letter, he speaks of the Word's role in bringing the universe to its completion in the Father:

> We worship the Word . . . as the Creator and Maker coming to be in a creature so that, by granting freedom to all in himself, he may present the world to the Father and give peace to all, in heaven and on earth.[33]

Through the incarnation of the Word, human beings are forgiven, deified, and adopted as beloved sons and daughters. The rest of the creaturely world is to be transformed in Christ in its own proper ways. As the later Greek theological tradition makes clear, there are distinctions between creatures in their way of participation in trinitarian life—they participate in the divine Communion according to their own proper capacity and their own proper nature. But in ways appropriate to each creature, the whole creation is to participate through the Word, in the Spirit, in the divine life of the Trinity. The receptivity of a human being is different from that of a tree. But both participate of the Word in the Spirit. And, of course, the way a saint participates of the Word in the Spirit is different from that of a sinner. In his reflections on this theme in Gregory Palamas, Orthodox theologian Doru Costache notes:

> All are in God yet not all are equally receptive to God's presence. The Spirit shines wholly and continually to all creation yet the receptive capabilities vary from one being to another.[34]

---

[33] Athanasius of Alexandria, "Letter 40: To Adelphius 8" (Anatolios, *Athanasius*, 242).

[34] Doru Costache, "Experiencing the Divine Life: Levels of Participation in St Gregory Palamas's *On the Divine and Deifying Participation*," *Phronema* 26, no. 1 (2011): 9–25, at 17.

God is radically and fully present to each creature, but creatures participate of the Word in the Spirit in distinct ways because of their distinct forms of receptivity to divine life.

For Athanasius, then, both the creation of a world of creatures and their deifying fulfillment are seen as partaking of the Word in the Spirit. This is encapsulated in Athanasius's sentence that I have quoted earlier: "The Father creates and renews all things through the Son and in the Holy Spirit."[35] From the perspective of the cross and resurrection, Athanasius sees God's act of creating, sustaining, providing for, and governing a world of creatures as occurring through their partaking of the Word in the Spirit. He sees the incarnation as bringing about the deifying forgiveness, healing, and transformation of mortal human nature by participation in the Word, through the Spirit. And he believes that in its own way the natural world will be healed and glorified, participating with human beings in their deifying adoption as daughters and sons of God.

Athanasius's incarnational theology, then, relates to the wider natural world in two interrelated ways: On the one hand, starting from the cross of Christ, this theology sees the whole universe and every creature in it as existing only because it partakes of the Word in the Spirit. On the other hand, it sees the natural world, in its own proper ways, as newly enabled to participate with human beings in the deifying transformation that comes through the incarnation of the Word in the Spirit.

Conclusion

I find that Athanasius's trinitarian theology of deification offers a direction for a contemporary theology of the wider creation as participating in salvation in Christ. Of course, the rest of creation has not sinned and does not need forgiveness. But it has been damaged by human sin and is still in process, still "groaning" in the pain of giving birth (Rom 8:22). It is not yet brought to its fulfillment. Paul thinks of the wider creation as longing for the liberation that will

---

[35] *Ep. Ser.* 1.24 (Anatolios, *Athanasius*, 224).

come to it only through its participation in the resurrection (Rom 8:19-23). Something has begun in the risen Christ that will one day be true of the whole creation.

Karl Rahner encapsulates this tradition in a dense but rich sentence. He sees the resurrection of Jesus as being ontologically "the irreversible and embryonically final beginning of the glorification and divinization of the *whole* reality."[36] The teaching of the final fulfillment of the whole creation in Christ finds expression in the Second Vatican Council's *Lumen Gentium* (Dogmatic Constitution on the Church): "At that time, together with the human race, the universe itself, which is closely related to humanity and which through it attains its destiny, will be perfectly established in Christ (see Eph 1:10; Col 1:20; 2 Pet 3:10-13)" (LG 48).[37]

The fulfillment of the whole creation that is promised in the resurrection is beyond our limited imagination. This is not surprising, seeing we know so little about our own life beyond death. Our fulfillment in God is beyond our imagining. As Rahner has pointed out, we have no insider information about resurrection life.[38] What we have is the experience of God in Christ and the experience of the Spirit in the life of grace. And in Christ and his resurrection we have the unbreakable promise of God. The promise is of a future in God. God is this future, the "Absolute Future." This future of ourselves and the wider creation is announced and promised in Christ, but it is announced and promised precisely as hidden mystery, the coming of the incomprehensible God. It is revealed only as "the dawn and the approach of mystery as such."[39] Deification, then, is our future and that of our world in God, in the God beyond imagining.

This means, as Paul tells us, that "we hope for what we do not see" (Rom 8:25). Our hope for the natural world, and for ourselves,

---

[36] Karl Rahner, "Dogmatic Questions on Easter," *Theological Investigations* 4 (London: Darton, Longman & Todd, 1966), 129.

[37] *Vatican Council II: The Basic Sixteen Documents*, trans. Austin Flannery (Northport, NY: Costello Publishing, 1996), 72.

[38] Karl Rahner, "The Hermeneutics of Eschatological Assertions," *Theological Investigations* 4, 323–46.

[39] Ibid., 330.

is based not on what we can see or imagine, but on the Word of God. While we have no good imaginative picture of this transformed existence, we can hope that it includes in some way flowers and forests, kangaroos and dolphins, and the whole of evolutionary history. In their own distinct ways, all creatures are embraced by God in the incarnation of the Word and are destined to be transformed in Christ, in the Spirit, and thus find their fulfillment in their own way in the divine Communion.

Jesus tells us that not one sparrow is "forgotten before God" (Luke 12:6). God does not forget or abandon any creature but inscribes it eternally in the divine life. The sparrow that falls to the ground is not abandoned, but is gathered up and in some way brought to new life in Christ, in whom "creation itself will be set free from its bondage to decay" (Rom 8:21). The sparrow that falls to the ground is among the "all things" that are reconciled (Col 1:20), recapitulated (Eph 1:10), and made new (Rev 21:5) in the risen Christ. What we know is the promise of God that is given in the resurrection of the Word made flesh. We can hope that, in our participation in the communion of saints, we will share in God's delight in other animals within the abundance and beauty of creation brought to its fulfillment. In particular we may hope that the relationships we have with particular creatures, such as a beloved dog, do not end with death, but are taken into eternal life. Each sparrow and each dog exists because it partakes of the Word in the Spirit, participates in its own way in deification in Christ, and is eternally held and treasured in the life of the Trinity.

Chapter 3

Deep Incarnation

The Meaning of Incarnation
for the Rest of the Natural World

For many Christians, the great liturgical act of the year begins on Holy Saturday night with the lighting of the paschal candle from the Easter fire. The community then enters the dark church following the candle, the symbol of the risen Christ. Each person's candle is lit from the paschal candle, light spreads throughout the church, and the great *Exsultet* is sung. Then, illuminated by the light cast by the paschal candle enthroned high on its stand, all listen to the reading of the Scriptures, beginning with the story of the creation in the opening chapter of Genesis. They look back on the creation of the universe and the history of salvation and see it all illuminated by the light of the crucified and risen Christ.

I see this as a good image for the particular way in which Christian theology, exemplified in Athanasius, sees the creation of the natural world from the perspective of Christ. All things are created through the Word and all things are to be liberated and fulfilled in the Word made flesh. While the incarnation of the Word in the created humanity of Jesus is a completely unique event, it is an event not only for Jesus and not only for the whole of humanity but also for the whole interconnected biological and physical world.

In this chapter I will explore the theology of incarnation in a way that builds on a trajectory found in Athanasius's work but also goes beyond it. I will begin by examining Athanasius's view of Wisdom's ways of being present to creatures. This will lead to the contemporary idea of "deep incarnation," as a way of understanding incarnation within the context of an evolutionary world. Then I will explore two implications of this view of incarnation for today: the first, on God's eternal commitment to the natural world, in dialogue with Karl Rahner and Thomas Torrance; the second, on God's engagement with the particular, in dialogue with Sandra Schneiders and Niels Gregersen.

Wisdom's Ways of Being with Creatures Culminate in Incarnation

Throughout his work, Athanasius sees the presence and action of the Wisdom of God revealed in the goodness, order, and beauty of the natural world. The natural world speaks eloquently of divine Wisdom, its Creator, but humanity's capacity to hear is damaged by sin. However, God's love for creation is not defeated by sin. Out of divine generosity, the Wisdom of God, already manifest through all creation, now comes to be with creatures in a radical way, in the flesh. The presence of Wisdom reaches an unforeseeable culmination in bodily incarnation. In Jesus of Nazareth, in his life, death, and resurrection, the Wisdom of God is interiorly present to creaturely humanity, bringing forgiveness, healing, and liberation, overcoming death and transforming creaturely existence from within.

One of the several places where Athanasius spells out this line of thought is at the end of his *Second Oration against the Arians*. The context is his defense of the divinity of Wisdom—in response to opponents who see Wisdom as a creature on the basis of the text: "The LORD created me at the beginning of his work, the first of his acts of long ago" (Prov 8:22). Athanasius responds by distinguishing between the imprint and reflection of Wisdom that he sees as found in each creature, and Wisdom herself. While the imprint of Wisdom in creatures is certainly created, Wisdom herself is not. He writes:

> Therefore, the only-begotten and true Wisdom of God is the creator and maker of all things. For it says: "In wisdom you have made all things" and "the earth is filled with your creation" (Ps 104:24). But in order that creatures may not only be but also thrive in well-being, it pleased God to have his own Wisdom condescend to creatures. Therefore he placed in each and every creature and in the totality of creation a certain imprint (*typon*) and reflection of the Image of Wisdom, so that the things that

come into being may prove to be works that are wise and worthy of God. Just as our word is an image of the Word who is Son of God, so the wisdom that comes into being within us is an image of his Wisdom, in which we attain to knowledge and understanding. Thus we become recipients of the Creator-Wisdom, and through her we are able to know her Father.[1]

This is a rich text that says several important things. One is the idea I have already discussed, that God has God's own Wisdom "condescend," or come down to be with creatures in immediate presence, in the act of continuous creation—the Wisdom of God is immediately present to every creature as the very source of its existence. Another is that God places "in every creature and in the totality of creation" an imprint (*typon*) and reflection of Wisdom. Whales, koalas, eagles, ants, and human beings are all in their own distinct and interrelated ways reflections of divine Wisdom, bearing the mark of Wisdom. This tree I see from my window, then, not only exists from Creator-Wisdom but also in itself is a created reflection of Wisdom, bearing Wisdom's imprint. And the universe we know, the dynamic, expanding observable universe with its 100 billion galaxies, reflects divine Wisdom and bears Wisdom's character. And Earth, our fruitful, vulnerable home, with its evolutionary history, with its wonderfully diverse life forms and the seas, land, and atmosphere on which life depends, reflects the beauty of divine Wisdom and is marked by Wisdom.

A third insight concerns Wisdom's image in human beings. Athanasius sees our human experience of wisdom as an image of, and a participation in, divine Wisdom: "the wisdom that comes into being within us is an image of his Wisdom, in which we attain to knowledge and understanding."[2] In our experience of nature, in our

---

[1] Athanasius of Alexandria, *Orations against the Arians* (= *C. Ar.*) 2.78 in Anatolios, *Athanasius* (New York: Routledge, 2004), 171. Anatolios comments on his translation of personal pronouns in this text: "In Athanasius's Greek, the personal pronoun switches from feminine when the subject is Wisdom (*Sophia*) to masculine when the subject is the Word (*Logos*) or the human being (*anthropos*) which the Word became" (267, n. 173).

[2] Ibid., 171.

interpersonal relationships, in our pursuit of justice, in our search for truth and understanding, in our pondering of the Word, in moments of silence, we can find the image of Wisdom within ourselves. We are led to Creator-Wisdom herself, and in knowing her we can know the Father. It might be said today that the growing sense in the human community that we are responsible for the well-being of the global community of life on Earth is not only something stirred up by the life-giving Spirit but also is a participation in Holy Wisdom.

Athanasius notes, with Paul, the sad fact that in spite of God's attributes being evident in the creation since the beginning, human beings have over and over failed to glorify God and have instead worshipped false gods (Rom 1:19-21). In the light of this failure, God does not abandon humanity, but out of the abundance of divine generosity sends Wisdom to be with us in the flesh:

> For God willed to make himself known no longer as in previous times through the image and shadow of wisdom, which is in creatures, but has made the true Wisdom herself take flesh and become a mortal human being and endure the death of the cross, so that henceforth all those who put their faith in him may be saved. But it is the same Wisdom of God, who previously manifested herself, and her Father through herself, by means of her image in creatures—and thus is said to be "created"—but which later on, being Word, became flesh (John 1:14) as John said.[3]

In Jesus Christ, Wisdom herself takes flesh in our midst to secure the deification of human beings and with them, the fulfillment of the rest of creation in God. When Athanasius says that since the incarnation God no longer reveals God's self though the image of Wisdom in creation, I take him to mean that God is no longer revealed in this way *alone*, but is now revealed in the utter extravagance of Wisdom made flesh. After the incarnation, Wisdom is still revealed through her imprint and reflection in all creation. But now, in the light of Wisdom incarnate, we have all the more reason to celebrate Wisdom's presence in the icons of Wisdom all around us, in great trees, tiny wild flowers, threatened species, and human beings.

[3] *C. Ar.* 2.81 (Anatolios, *Athanasius*, 174).

Deep Incarnation

In what follows, I will seek to build on Athanasius's insight into the incarnation as the culmination of the Wisdom/Word's presence and action with the contemporary concept of "deep incarnation." This idea was introduced by Danish theologian Niels Gregersen and has since been taken up by other theologians in different ways, including Elizabeth Johnson and Celia Deane Drummond as well as myself.[4] Gregersen writes that "the incarnation of God in Christ can be understood as a radical or 'deep' incarnation, that is, an incarnation into the very tissue of biological existence, and system of nature."[5] In the incarnation, God in Christ enters into biological life to be with the evolving creation in a radically new way. Gregersen understands the cross as God's identification with creation in its evolutionary emergence, and as a microcosm of God's redemptive presence to all creatures in their suffering and death.

I think that Athanasius's theology of the incarnation can ground a contemporary understanding that Wisdom's embrace of flesh involves not just the humanity of Jesus, and not just the wider human community, but the whole natural world in its evolution and its groaning. This is not at all to suggest that God is incarnate in all things in the same way God is incarnate in Christ. Such a view would undermine the newness and the absolute gratuity of the incarnation and ultimately reduce incarnation to creation. The proposal is, rather, that the absolutely gratuitous and unparalleled event of the incarnation of the Word in Jesus of Nazareth is an event that not only uniquely unites the Word of God with Jesus'

---

[4] Niels Henrik Gregersen, "The Cross of Christ in an Evolutionary World," *Dialog: A Journal of Theology* 40 (2001): 205; Niels Henrik Gregersen, "Deep Incarnation: Why Evolutionary Continuity Matters in Christology," *Toronto Theological Journal* 26, no. 2 (2010): 173–88; Denis Edwards, *Ecology at the Heart of Faith* (Maryknoll, NY: Orbis Books, 2006), 58–64; Elizabeth Johnson, "An Earthy Christology: 'For God so Loved the Cosmos,'" *America* 200, no. 12 (April 13, 2009): 27–30; Celia Deane-Drummond, *Christ and Evolution: Wonder and Wisdom* (Minneapolis, MN: Fortress Press, 2009), 128–55. For a wide range of views on this topic see *Incarnation: On the Scope and Depth of Christology*, ed. Niels Henrik Gregersen, forthcoming from Fortress Press.

[5] Gregersen, "The Cross of Christ," 205.

own creaturely reality but also, in and through this union, unites the Word in a gratuitous and new way to all biological and material reality, in all its unfinished, evolutionary nature. As I have explored in the last chapter, it involves the transformation of the whole creation in Christ.

The concept of deep incarnation is meant to reflect not only the kind of life-changing insight into the incarnation that Athanasius represents but also some of the new insights that come to us today from evolutionary biology. It proposes that, in the unique incarnation of the Word in Jesus of Nazareth, God embraces our wider humanity, and the wider community of life on Earth, and the whole evolutionary universe of which we are a part. Of course, the humanity of Jesus can now be understood as, like all of humanity, dependent upon the evolution of life from its microbial origins 3.7 billion years ago. In the bodily humanity of Jesus, God is made one with all the fruits of evolution by means of natural selection. The body of Jesus is made up of atoms produced in the nuclear furnaces of stars. It depends upon the cooperation of the billions of microbes that inhabit it. It exists only in interdependence with all the other organisms and all the systems that sustain life on Earth.

In a biological view, it makes no sense to think of one person's human flesh as an isolated reality. It can be understood only as interrelated and interdependent with the whole of life and all that supports life, including the atmosphere, the land, and the seas. It involves dependence on the creatures that have gone before us in evolutionary history, and our ecological interdependence involves all the processes by which life has evolved. Reflecting on the incarnation in the light of our evolutionary heritage, and the crisis of life on our planet, we are led to a deeper appropriation of the meaning of *God-with-us* in Christ, as a theology of *God-with-all-living-things*. In the Word made flesh, God embraces the whole of finite creaturely existence from within. The incarnation is God-with-us in the "very tissue of biological existence" and in the systems of the natural world. One of the startling implications of the Christian view of the depths of the incarnation is that it is a claim about a God who eternally binds God's self to flesh and to matter.

Forever a God of Flesh and Matter

A consequence of a thoroughly incarnational theology is that God is understood as becoming *forever* a God of matter and flesh. This is the meaning of the Christian doctrines of the resurrection and the ascension. The Word is made flesh, and matter and flesh are irrevocably taken to God and embedded in the divine Trinity. The incarnation and its culmination in the resurrection and ascension of the crucified Jesus mean that the Word of God is forever matter, forever flesh, forever a creature, forever part of a universe of creatures, but part of all of this that is now radically transfigured. As the firstborn of the new creation the risen Christ is the beginning of the deifying transformation of the whole universe of creatures in God.

This is a theme that can be found running through many of Karl Rahner's writings. In 1950, for example, he published a short article entitled "A Faith that Loves the Earth," in which he ponders the meaning of Christ's resurrection. He sees Jesus as descending in his death to the "heart of the earth" (Matt 12:40), entering fully into the place of creaturely impermanence and death, "God's heart at the center of the world." In the resurrection, he does not abandon this earthly reality. He is raised precisely in the body, and thus as the beginning of the liberating and life-giving transformation of creaturely reality: "No, he is risen in his body. That means: He has begun to transfigure this world into himself; he has accepted this world forever; he has been born anew as a child of this earth, but of an earth that is transfigured, freed, unlimited, an earth that in him will last forever and is delivered from death and impermanence for good."[6]

The risen Christ is still part of Earth, connected to Earth's nature and destiny. The new forces of a transfigured world are at work in him, and they are conquering impermanence, death, and sin at their core. While we continue to experience suffering and sin in the world, Christian faith holds that they have actually been defeated

---

[6] Karl Rahner, "A Faith that Loves the Earth," in *The Mystical Way in Everyday Life: Sermons, Essays and Prayers: Karl Rahner, SJ*, ed. Annemarie S. Kidder (Maryknoll, NY: Orbis Books, 2010), 55.

deep down at their very source. At this level there is no longer any distance between God and the world. Christ is already at the heart of the nameless yearning of all creatures that are waiting to participate in the transfiguration of Christ's body. Christ is at the heart of Earth's history, at the heart not only of love and generosity but also of tears, death, defeats, and even sin, as radical mercy, unbounded love, and the promise of life.

As Christians we do not, or at least should not, think that we need to escape from matter and flesh in order to love God. We are called to love the things of Earth and God *together*, because in the resurrection of Jesus "God has shown that he has adopted the earth forever."[7] Tertullian said long ago that the flesh is the connecting point, the hinge, of salvation: *Caro cardo salutis*. The Christian claim is not that we find God by going to God in the transcendent spiritual world beyond, but that God has come to us in the flesh. And it is as creatures of flesh that we are transformed in Christ. Since the incarnation, we know that "Mother Earth has brought forth only creatures that will be transfigured, for his resurrection is the beginning of the resurrection of all flesh."[8]

In another essay, a much later one, Rahner asks himself the question: what is truly specific to the Christian view of God? The answer he finds is precisely the idea that God gives God's very self to creation in the Word made flesh and in the Spirit poured out. What is truly characteristic of Christianity, Rahner claims, is that while maintaining the radical distinction between God and the world, it understands God's self-giving as the very "core" of the world's reality and sees the world as truly the "fate" of God. He writes:

> God is not merely the one who as creator establishes a world distant from himself as something different, but rather he is the one who gives himself away to this world and who has his own fate in and with this world. God is not only himself the giver, but he is also the gift. For a pantheistic understanding of existence this statement may be completely obvious. For a

---

[7] Ibid., 58.
[8] Ibid., 55.

Christian understanding of God, in which God and the world are not fused but remain separate for all eternity, this is the most tremendous statement that can be made about God at all. Only when this statement is made, when, within a concept of God that makes a radical distinction between God and the world, God himself is still the very core of the world's reality and the world is truly the fate of God himself, only then is the concept of God attained that is truly Christian.[9]

The idea that God is the "core" of the world's reality challenges many assumptions that we Christians tend to make about the natural world around us. Even more confronting is Rahner's claim that this material and fleshly world, this world of creatures in evolutionary process, is the "fate" of God. The Word is made flesh and flesh is taken to God irrevocably. The resurrection and ascension of the crucified Jesus means that the Word of God is forever part of evolutionary history on this planet and forever part of a universe of creatures. In creation, incarnation, and its culmination in resurrection, God commits God's self to this world, to this universe and its creatures, and does this eternally. In the risen Jesus, part of the biological community of Earth is already forever with God as the sign and promise of the future of all things in God.

Rahner, a Jesuit theologian in the Roman Catholic tradition who was deeply involved in rethinking the heritage of Aquinas, found the largely juridical interpretation of the incarnation in Western theology inadequate and that he needed to turn to the Eastern tradition of deification for a richer theology of the meaning of Christ.[10] This turn to the East also occurs in the work of some Protestant theologians. Thomas Torrance, a minister of the Church of Scotland who was strongly influenced by Calvin and his own teacher, Karl Barth, represents a different theological tradition to Rahner. But Torrance, too, draws on Eastern patristic theology, and his trinitarian theology

---

[9] Karl Rahner, "The Specific Character of the Christian Concept of God," *Theological Investigations* 21 (New York: Crossroad, 1988), 191.

[10] Karl Rahner, "Dogmatic Questions on Easter," *Theological Investigations* 4 (London: Darton, Longman & Todd, 1966), 122–26.

builds explicitly on Athanasius. Torrance writes of the meaning of incarnation:

> Through his cross and resurrection the incarnate Saviour penetrated into the ontological depths of creation where in death created being borders upon nonbeing, and set it upon a new basis, that of Grace in the triumph of God's holy Love in what the Bible speaks of as a new heaven and a new earth.[11]

In the Christ event, the new creation is inaugurated in the midst of the old. This is an event that embraces all times:

> The incarnation was not just a transient episode in the interaction of God with the world, but has taken place once-for-all in a way that reaches backward through time and forward through time, from the end to the beginning and from the beginning to the end.[12]

The Creator of the universe of creatures has once for all become incarnate in it. The incarnation means that the whole universe is brought to share in the freedom of the Creator in the differentiated way appropriate to each creature. God irreversibly holds the created universe to God's own being. Torrance sees the incarnation as meaning that "God has decisively bound himself to the created universe and the created universe to himself, with such an unbreakable bond that the Christian hope of redemption and recreation extends not just to human beings but to the universe as a whole."[13]

The claim made by Torrance and by Rahner is a large one, but it is one that I see as building authentically on the Christian conviction, articulated so powerfully in the work of Athanasius, that Jesus Christ is the eternal Wisdom/Word of God who is made flesh that we, and with us other creatures, might be saved and deified in him. His incarnation constitutes an unbreakable bond with the whole creation, to all creatures of all times and all places.

---

[11] Thomas Torrance, *The Christian Doctrine of God: One Being Three Persons* (Edinburgh: T & T Clark, 1996), 214.

[12] Ibid., 216.

[13] Ibid., 244.

In creation, incarnation, and in its culmination in resurrection, God commits God's very self to this world, to this universe and its creatures, and does this eternally. In the risen Jesus, part of this biological community of Earth, of this evolutionary history, and of this material universe is already transfigured in God, as the sign and promise of the deifying fulfillment and transformation of all things in God. In the Creator Spirit, this same divine Wisdom is already at work in the whole universe of creatures, bringing them to their liberation and completion in God.

The Scandal of God's Engagement with the Particular

The very particularity of the incarnation of God in one human being can seem far too specific and limited to offer meaning for the whole of reality. This is sometimes called the "scandal of particularity." This scandal is greatly exacerbated by exposure to contemporary science. When cosmologists tell us that our observable universe has been expanding and evolving over the last 13.75 billion years, that our Milky Way Galaxy is made up of something like 200 billion stars more or less like our Sun, and that there are perhaps 100 billion galaxies in the observable universe, then Christian claims about the incarnation can appear too confined, too specific, and too concrete. When we think, in addition, not only of the pluralism of religions on our planet but also of the possibility of intelligent and religious life on other planets, then it becomes easy to understand that some Christians have backed away from universal claims for Jesus Christ and, like Wesley Wildman, opted for what he calls more "modest" christologies.[14]

But as biblical scholar Sandra Schneiders has said, everything ultimately depends on the kind of interpretation we make of Jesus. One possible interpretation is "to *reduce* Jesus to his particularity as a first-century Jewish male who lived a short life in one small country, was executed, and is now a figure of history whom we admire and even imitate but with whom we cannot relate personally and whom

[14] Wesley Wildman, *Fidelity with Plausibility: Modest Christologies in the Twentieth Century* (Albany: State University of New York Press, 1998).

we must not universalise."[15] Once Jesus is understood simply as a great human being and moral example, this leads inevitably to the conclusion that he cannot have meaning for the whole of human history, let alone for the universe that science puts before us. Then, Schneiders says, he becomes "substantively irrelevant for the scientifically and interreligiously enlightened contemporary person."[16]

A second possible way of interpreting Jesus, Schneiders says, is "to take utterly seriously the faith of the Church that Jesus *is the Wisdom of God incarnate*."[17] She points to the biblical and patristic tradition, where Wisdom is seen as the immanence of the transcendent God who is present and active in all creation. Wisdom creates, sustains, and brings the universe of creatures to completion and wholeness. It is this Wisdom that the tradition sees as made flesh in Jesus of Nazareth. Schneiders comments that such an interpretation does not imprison or restrict the infinity of God, but rather "focuses" infinity so that in our finitude we can encounter and relate to the absolute mystery. Obviously, if one accepts that in the particular created humanity of Jesus, the eternal, transcendent, creative, Wisdom of God really does becomes flesh, then something is being said in Jesus Christ that can have the deepest meaning for human beings and their history, for Earth and all its other creatures, and for the whole universe.

One of the legacies of modernity, I believe, is the sometimes unnoticed tendency to simply assume the first of these interpretations—to operate as if Jesus is simply a great prophetic figure. This can happen, I think, with the best of intentions, in genuine attempts to communicate the gospel in a secular world. But I wish to suggest that we need to resist this tendency and claim afresh the great tradition of the divine Wisdom of God present to us in the particularity of the human face of Jesus.

A fundamental implication of accepting the full biblical and patristic interpretation of the incarnation, as Schneiders observes, is

---

[15] Sandra Schneiders, "The Word in the World," *Pacifica* 23 (2010): 263.
[16] Ibid.
[17] Ibid.

that *particularity* is revealed as infinitely precious.[18] This can mean that the God of the universe is revealed not only in the particularity of Jesus but also in this laughing kookaburra, this beautiful flowering eucalyptus tree, this vulnerable human person before me. God is present in, and revealed in, the finite, the ordinary, as well as in the wonders of our planet and our universe.

Divine engagement with the particular extends to every aspect of God's creative act in the emergence of the universe and the evolution of life on Earth. In his theological engagement with the sciences of complexity, Niels Gregersen has proposed that the self-organization that science discovers at work in complex processes be interpreted theologically as the gift of the Creator. The benevolence and generosity of God is such that God bestows on creatures the capacity to make themselves. God designs creation for self-organization: "God's design of the *world as a whole* favors the emergence of autonomous processes in the *particular course* of evolution, a course at once constrained and propagated by a built-in propensity towards complexification."[19] From a theological perspective, Gregersen sees the effectiveness of self-organization as exemplifying a principle of grace written into the structure of the natural world.

In a later work, Gregersen reflects on how we might think of the Creator as acting in these emergent processes. He points out that in a self-organizing system like a cell there is a constant "rewiring" that occurs in interaction with the environment, involving an enormous number of steps. These steps are not covered by any one scientific law, but require a variety of interacting scientific explanations.[20] The kind of complexity discovered in self-organizing systems at the heart of the natural world suggests a theological conclusion to Gregersen. If God is engaged with every aspect of ongoing creation, then God's

---

[18] Ibid., 262.

[19] Niels Henrik Gregersen, "From Anthropic Design to Self-Organized Complexity," in *From Complexity to Life: On the Emergence of Life and Meaning*, ed. Niels Henrik Gregersen (Oxford, UK: Oxford University Press, 2003), 207–8.

[20] Niels Henrik Gregersen, "Laws of Physics, Principles of Self-Organization, and Natural Capacities: On Explaining a Self-Organizing World," in *Creation: Law and Probability*, ed. Fraser Watts (Aldershot, Hampshire, 2008), 97–98.

engagement is not simply of a general kind. It is better thought of as special divine action that engages with the particulars: "For if God is not in the particulars, God is not in the whole of reality either."[21] Gregersen suggests that we can see God as involved in a kenotic way in all the particular details of self-organizing creation, "giving room—from moment to moment, from event to event—to the explorative capacities of God's creatures."[22]

In the evolution of the universe and in the emergence of life on Earth, divine action involves the historical, the unpredictable, and the specific. It involves the specific details of the lives of all living creatures, and in a unique interpersonal way, of human beings. This means we should take seriously the claim of Jesus when he says of sparrows that "not one of them is forgotten in God's sight," and when he makes a promise of God's providential care for humans: "But even the hairs on your head are all counted" (Luke 12:6-7).

In giving God's self to us in the Word made flesh and in the gift of the Holy Spirit, God is revealed as a God who is infinitely relational, a Communion of love that embraces difference. This God is creatively present through the Word and in the Spirit with each creature in all its specificity, and accompanies each in love. The incarnation of the Word in the Spirit and its fulfillment in resurrection constitutes an unbreakable commitment by God to bring the whole natural world to its proper, transfigured, deifying fulfillment in the divine life of the Trinity.

---

[21] Ibid., 98.
[22] Ibid.

Part Two

Darwin's "Tangled Bank"

The final paragraph of Charles Darwin's great work *On the Origin of Species* (1859) begins with his famous image of the tangled bank. It provides a brief summary of his view of the natural selection that gives rise to the variety of creatures that inhabit the bank:

> It is interesting to contemplate a tangled bank, clothed with many plants of many kinds, with birds singing on the bushes, with various insects flitting about, and with worms crawling through the damp earth, and to reflect that these elaborately constructed forms, so different from each other, and dependent upon each other in so complex a manner, have all been produced by laws acting around us. These laws, taken in the largest sense, being Growth with reproduction; Inheritance which is almost implied by reproduction; Variability from the indirect and direct action of the conditions of life, and from use and disuse; a Ratio of Increase so high as to lead to a Struggle for Life, and as a consequence to Natural Selection, entailing Divergence of Character and the Extinction of less improved forms. Thus, from the war of nature, from famine and death, the most exalted object which we are capable of conceiving, namely, the production of the higher animals, directly follows.[1]

Darwin recognizes that some might see this as a diminished vision of life. But he insists that, on the contrary, there is "grandeur" in

---

[1] Charles Darwin, *The Origin of Species by Means of Natural Selection of Favoured Races in the Struggle for Life* (New York: Signet Classics, 2003), 459.

this view of life, in which "from so simple a beginning endless forms most beautiful and most wonderful have been, and are being evolved."[2] Variations in inheritance are at the heart of Darwin's theory, variations that enable some offspring to better survive and reproduce in an environment. Unknown to Darwin, the Austrian monk Gregor Mendel (1822–84) had already begun to explore the genetic basis for such variations.

Biological Evolution Today

In the twentieth century, Darwin's theory of natural selection was combined with genetics to form the "Modern Synthesis" or "neo-Darwinism." There has been an astonishing evolution of Darwinism itself, above all, in the truly spectacular advances that have occurred in molecular biology, exemplified in the discovery of the structure of DNA by James Watson and Francis Crick in 1953 and in the human genome project at the turn of the century. In the first part of the twenty-first century there have been major developments in the biological explanation of the emergence not only of cooperation but also of language, ethics, economics, and religion.

Biologist Jerry Coyne does not appear to be a supporter of a dialogue between science and theology—his view of theology seems based on fundamentalist Christianity. But in a recent widely read book, he offers a summary of the modern theory of biological evolution that I find helpful as a theologian engaging with evolution: "Life on earth evolved gradually beginning with one primitive species—perhaps a self-replicating molecule—that lived more than 3.5 billion years ago; it then branched out over time, throwing off many new and diverse species; and the mechanism for most (but not all) of evolutionary change is natural selection."[3] Coyne goes on to discuss six elements that are encapsulated in this description, which I summarize briefly here:

[2] Ibid.

[3] Jerry A. Coyne, Why Evolution Is True (London: Penguin Books, 2009), 3.

1. *Evolution:* Species change and can evolve because of changes in DNA that are caused by genetic mutations.

2. *Gradualism:* It takes many generations to produce substantial change, with different species evolving at different rates, and always in relationship to environmental pressures.

3. *Speciation:* A splitting occurs when groups evolve to the point where they can no longer interbreed, or exchange genes, resulting in the more than ten million species of today.

4. *Common Ancestry:* Ancestry, the other side of speciation, can now be studied genetically, on the basis that species with more similarities in their DNA will have a more recent common ancestor.

5. *Natural Selection:* Individuals of a species vary genetically in their ability to survive and reproduce in their environment, making evolution inevitable, thus accounting for the appearance of design in nature.

6. *Processes other than Natural Selection:* Some evolution occurs through processes such as genetic drift—the change in the frequency of a gene variant (allele) in a population due to the random combination of genes passed from parents to offspring.[4]

Darwin's Gift to Theology

Theologian John Haught, who has long engaged creatively with the theory of evolution, writes of Darwin's work as a "gift to theology." Haught is well aware of the complex challenges involved in receiving Darwin's gift. First of all, it is clear that accepting Darwin's theory means the end of the simple argument from the appearance of design in nature to God as the great Designer. What appears to be intentional design in the natural world can now be explained as the product of natural selection. But Haught finds this not so much a problem as a helpful clarification for theology. He points out that

---

[4] Ibid., 3–14.

a theology grounded in an adequate philosophical view of the God-world relationship, and on Christian faith in God's self-revelation in Christ, does not need the early nineteenth-century Grand Designer kind of argument.

However, Darwin presents a further, more demanding, problem to theology: the problem of pain and death as built into the natural world. As Haught points out, Darwin himself understood natural selection not only as an effective mechanism but also as a pitiless one. It is clear that the age-old problem of evil is intensified when one considers the costs involved in the 3.7 billion years of evolutionary history—think of all the pain, predation, death, and extinctions that have occurred. Haught sees Darwin as challenging theology to think again about its view of God, above all, by plumbing the depths of its own insight into the humble, self-emptying God of the cross.[5] He says of Darwin that "his revolutionary and ragged view of life will eventually have to be taken into account in any realistic theological understanding of God, the natural world, life, human identity, morality, sin, death, redemption, and the meaning of our lives."[6] This is quite a list, and one that could easily be continued. After Darwin, theology cannot plausibly be the same as before, any more than it could simply continue on as before after Galileo. In one way or another, "Darwin has altered our understanding of almost everything that concerns theology."[7]

---

[5] John F. Haught, *God after Darwin: A Theology of Evolution*, 2nd ed. (Boulder, CO: Westview Press, 2008), 49–60. Evolutionary biologist Francisco Ayala also takes up the image of the gift in *Darwin's Gift: To Science and Religion* (Washington, DC: Joseph Henry Press, 2007). More generally, see Celia Deane-Drummond, *Christ and Evolution: Wonder and Wisdom* (Minneapolis, MN: Fortress Press, 2009); Ted Peters and Martinez Hewlett, *Theological and Scientific Commentary on Darwin's Origin of Species* (Nashville, TN: Abingdon Press, 2008); F. LeRon Shults, *Christology and Science* (Aldershot, UK: Ashgate, 2008); Cynthia Crysdale and Neil Ormerod, *Creator God, Evolving World* (Minneapolis, MN: Fortress Press, 2013).

[6] John F. Haught, *Making Sense of Evolution: Darwin, God, and the Drama of Life* (Louisville, KY: Westminster John Knox Press, 2010), xv.

[7] Ibid.

With Haught, I see the insights of evolutionary biology as a gift to theology, an invitation to think again about God, Jesus Christ, and every other aspect of theology. What I am attempting here can only be a partial response, of course. The trinitarian theology of Athanasius, I have proposed, can be a resource for rethinking the relationship between God and the whole natural world. In the light of Darwin, and in the light of the crisis of life on Earth, however, far more needs to be said. We face the requirement for theological developments that Athanasius never needed to consider. My proposal, however, is that the kind of dynamic trinitarian theology found in Athanasius offers a deep and strong foundational structure on which to explore new, desperately needed theological developments in today's evolutionary and ecological context.

In this second section of the book, I will take up five such developments: In the first, I will suggest transforming Athanasius's trinitarian theology of the Word and Spirit into a theology of the Word as Attractor and the Spirit as the Energy of Love in evolutionary emergence. In the second development I will focus on the suffering of God for creation. In the third, I will discuss the humble God in relation to the autonomy of evolutionary processes. In the fourth, I will reflect on church teaching about the human soul in relation to neuroscience's insights into emergence of the human brain. Finally, I will explore the connection between evolution and the doctrines of grace and original sin.

Chapter 4

The Attractor
and the Energy of Love
Trinity in Evolutionary Context

In Athanasius's view, stars, rocks, fish, birds, and humans do not have in themselves the reason for their own existence. They exist because they are held in being by the triune God. The Source of All continuously creates all things out of nothing through the Word in the Spirit. The indwelling Holy Spirit enables each entity to participate in the creative Word of God. The Word of God is immediately present to each entity, at every moment enabling it to exist and to interact in a community of creatures.

In a more metaphysical style, Thomas Aquinas develops this theology of creation. He sees God as always and everywhere conferring existence on all things. He calls this continuous relationship by which God enables things to exist "primary causality." He sees all the interactions of creatures, all the empirical processes we can observe, and all that the sciences we study as "secondary causes." He insists that these creaturely interactions have their own integrity. God does not operate in competition with creaturely processes, does not overturn the laws of nature, but acts creatively and providentially in and through the whole process.

I believe that this theological vision retains its importance today. It makes it clear that we look to science to explain our evolutionary history. There is no need for a "god of the gaps." There is no need for anything like "Intelligent Design." God is not to be thought of as occasionally or regularly replacing creaturely causes. God acts in and through the whole, and every part of it, conferring on creatures their existence and the capacity to act according to their own proper character. As the classical theology of creation puts it, God creates all things out of nothing, continually sustaining them in their being (*conservatio*) and enabling their actions (*concursus*).

Aquinas's is a rich, robust theology of creation that remains deeply meaningful. But biology after Darwin presents us with a vision of all life on Earth as evolving from its microbial origin 3.7 billion years ago, and cosmology after Einstein and Hubble sees our observable universe as expanding and evolving from a tiny, hot, dense state 13.75 billion years ago. The creation theology of thinkers like Athanasius and Aquinas requires development if it is to speak to the dynamic, evolutionary worldview that we receive from the sciences. In this new context, the relationship of creation will need to be seen as conferring on creatures not only existence and the capacity to act but also the capacity for evolutionary emergence.

The evolutionary nature of the universe invites us to think again about God. It suggests that we need a theology in which the triune God is seen as the source not only of the existence of the natural world but also of its fruitful creativity and capacity for novelty. We need a theology of a God who, like Mother Carey in Charles Kingsley's *The Water Babies*, "can make things make themselves."[1] One of the "gifts" that evolutionary biology offers to theology, I believe, is precisely that it encourages us to think more explicitly of a God who creates in such a way that creatures participate in the process. God seems to be a God who delights in the emergence of creaturely reality through increasing complexity.

In this chapter, building on the foundation of Athanasius, I am proposing a development of trinitarian theology in the light of an evolutionary worldview, in which the Holy Spirit is conceived as the immanent Energy of Love at work in evolutionary emergence and the Word of God is understood as the divine Attractor. In taking this approach, I hold to the traditional view that the Trinity's actions with regard to creation are one and undivided but also to the view that something specific can be said of each person in the unity of their one act.[2]

---

[1] Charles Kingsley, *The Water-Babies: A Fairy Tale for a Land-Baby* (New York: Hart Publishing Company, 1977), 255–56.

[2] I have argued this in some detail in *Breath of Life: A Theology of the Creator Spirit* (Maryknoll, NY: Orbis Books, 2004).

The Spirit as the Energy of Love in Evolutionary Emergence

Karl Rahner has made an important contribution to theology in an evolutionary context with his idea of the Creator as enabling the active "self-transcendence" of creatures.[3] The broader context for this insight is Rahner's vision of creation as a divine act of self-bestowing love, which is always directed toward the incarnation. A central effect of God's self-giving in continuous creation is that it enables creatures to become something new, to transcend themselves, through all the processes of evolutionary emergence studied in the sciences. This capacity is God-given in the ongoing relationship of creation that exists between the Creator and all creatures. But it gives to creatures themselves the capacity to become. This self-transcendence is at work in the processes of emergence that cosmologists find in the observable universe and in the evolutionary processes that biologists discover at work in life on Earth, above all in the transitions from inert matter to life and from life to the human.

Rahner's insight transforms the classical theology of creation and enables it to function in a new, evolutionary era. I am proposing two modifications of Rahner's theology. First, instead of the philosophical language of "self-transcendence," I will use terminology more closely connected to the sciences, speaking of the natural world's God-given "capacity for evolutionary emergence." God's creative act can then be understood as calling forth a universe of creatures and conferring on this creaturely world the capacity for its own evolutionary emergence. Second, where Rahner often speaks of God in philosophical terms, I will employ trinitarian categories, envisioning the Spirit as the life-giving Energy of Love that enables the evolutionary emergence of our universe and all its creatures—"When you send forth your spirit, they are created; and you renew the face of the ground" (Ps 104:30).

The proposal, then, is that the Spirit of God is to be seen as immanently present to all the entities of our universe, enabling creatures to exist, to interact, and emerge into the new by means of the

---

[3] Karl Rahner, *Foundations of Christian Faith: An Introduction to the Idea of Christianity* (New York: Seabury Press, 1978), 183–87.

laws of nature and the processes discussed in the natural sciences. The capacity for emergence, for increase in complexity through the self-organizational processes that are at work throughout the universe, and for the evolution of life on Earth by means of natural selection, is interior to creaturely reality. It belongs to the natural world. The capacity for emergence comes from within. At the empirical level of science, the emergence of the new is completely open to explanation at the scientific level. But, I am proposing, this capacity is Spirit-given, the gift of the Spirit's empowering, life-giving presence to creatures in the relationship of continuous creation.

The evolution of the observable universe and its creatures requires explanation at more than one level. It cannot be explained simply at the level of theology or by a literalist reading of Scripture, as some creationist Christians suppose. Nor can it be explained fully at the level of empirical science as some evolutionary naturalists claim. As John Haught points out, evolutionary emergence is one of very many realities we encounter that require more than one layer of explanation. He offers the example of a page of a book. One layer of explanation for the page is that it exists because a printing press stamps black ink on white paper. At a deeper level, the page exists because an author is trying to communicate something to readers. At a third level the page exists because a publishing house wants to add a particular book to its publishing lists. These three layers of explanation are not competing. They have some independence of each other, yet each is needed to account for the page. In a similar way, Haught suggests, explanations for evolutionary emergence can be multilayered, and are not necessarily to be thought of as in competition with one another: "The causal effectiveness of natural selection at one level of explanation is perfectly compatible with that of divine creativity operating at a deeper level."[4]

The Holy Spirit's presence and action operates at a level that is beyond the empirical. In *A Brief History of Time*, physicist Stephen Hawking asks a well-known question:

---

[4] John Haught, *Making Sense of Evolution: Darwin, God and the Drama of Life* (Louisville, KY: Westminster John Knox Press, 2010), 26.

Even if there is only one possible unified theory, it is just a set of rules and equations. *What is it that breathes fire into the equations* and makes a universe for them to describe? The usual approach of science of constructing a mathematical model cannot answer the questions of why there should be a universe for the model to describe. Why does the universe go to all the bother of existing?[5]

Hawking is asking a metaphysical and theological question, one that cannot be answered by the methods of empirical science. Even when, in his later work, Hawking seeks to account for the origin of the universe with a theory that it arises from random fluctuations in the quantum vacuum, the question remains: What is it that accounts for the very existence of the possibilities built into the quantum vacuum and what is it that breathes fire into the emergence of the universe and all its creatures?

Christian theology understands God as breathing the fire of the Spirit into the equations. The Breath of Life of the Bible can be seen as breathing life into the laws of nature and into all the natural processes by which the universe and all of life on Earth emerges. The Spirit can be imaged in the blazing flame that confronts Moses in the burning bush (Exod 3:2) and the fire of Pentecost (Acts 2:3), in the whirlwind that is the place of God for Job (Job 38:1), in the wild wind that blows where it wills of John (John 3:8), and in "the sound of sheer silence" that is the sign of the divine presence for Elijah (1 Kgs 19:12). The Spirit of God is symbolized in physical form as "living water" (Jer 2:13; John 4:14; 7:39), and in animal form as "like a dove" (Mark 1:10). Very often in the Scriptures the Spirit is the Breath of Life (Gen 2:7; Job 34:14-15; Ps 33:6; Ps 104:27-30; Ezek 37:3-10), and in the Creed the Spirit is the Life-Giver.

In summary, for biblical faith the Spirit is the "vivifying and energizing power of God" immanently present to all things.[6] This

---

[5] Stephen Hawking, *A Brief History of Time: From the Big Bang to Black Holes* (New York: Bantam, 1988), 174.

[6] John L. McKenzie, "Aspects of Old Testament Thought," in *The New Jerome Biblical Commentary*, ed. Raymond Brown, Joseph Fitzmyer, and Roland Murphy (Boston, MA: Pearson, 1989), 1291.

life-giving, energizing power of the Spirit is poured out on the Christian community at Pentecost, constituting it as the church of Jesus Christ. This same Holy Spirit is the power of divine love at work in us by grace: "God's love has been poured into our hearts through the Holy Spirit that has been given to us" (Rom 5:5).

The Spirit who is love at work in us is also the divine energy of love at work throughout the universe enabling its evolutionary unfolding. As I have pointed out, for Athanasius, the Holy Spirit is the energizing presence of God to all creatures. In the relationship of creation, the Spirit is the bond of communion between each entity and the eternal Word. In an evolutionary context, this relationship is understood as enabling not only the existence of entities but also their evolutionary becoming. Without this relationship there is no existence and no emergence. The Spirit is the indwelling Energy of Love that enables both.

Science rightly sees the emergence of the new (as when life first appears in a lifeless universe) as something it should seek to explain at the empirical level, without theological interference. But this emergence is also rightly seen by theology as God-given. For one who holds to both science and Christian faith, it can be understood as the result of the indwelling presence of the life-giving Spirit, enabling the new to emerge from within the natural world itself, by means of the processes, relationships, and causal connections studied in the natural sciences.

Emergence is a creaturely reality, but it occurs because of God's ongoing creative presence in the Spirit. It is this trinitarian presence in the Spirit that enables creatures to exist, to interact, and to emerge into the wonderfully new. The Spirit, then, is the Energy of Love in the emergence of the first particles of the early universe, in the birth of galaxies and stars, in the development of our solar system around the young Sun, in the origin of the first microbial life on Earth, in the flourishing of life in all its diversity, and in the emergence of humans with their capacity for consciousness and interpersonal love.

Walter Kasper sees the Holy Spirit as divine love in person at work in creation. The Spirit is the free overflow of dynamic triune love in our world. In a particular way, he understands the Spirit as source of the new in both the evolutionary unfolding of the universe of creatures and in human culture and life. In Kasper's thought, the Holy Spirit who enables creatures to participate in God's being thus also constantly enables them to become something new:

> The Holy Spirit is the internal (in God) presupposition of the communicability of God outside of himself. But the Spirit is also the source of movement and life in the created world. Whenever something new arises, whenever life is awakened and reality reaches ecstatically beyond itself, in all seeking and striving, in every ferment and birth, and even more in the beauty of creation, something of the being and activity of God's Spirit is manifested.[7]

Science puts before us a universe that comes to be in patterns of emergent relationships that nest within one another at many levels, including particles, cells, organisms, populations, ecological interactions, the planetary community, the solar system, the Milky Way Galaxy, and the observable universe itself. It suggests the possibility that extraterrestrial forms of life may exist not only on planets now being discovered in the Milky Way Galaxy but also in the billions of other galaxies that make up our universe. Theology sees the Spirit of God, the Energy of Love, as creatively and lovingly present, enabling these entities to interact and to become in an interrelational world. In a theological vision, each creature exists in such an interrelational world because the triune God holds it in love, and because the life-giving Spirit of God dwells in it (Wis 11:24–12:1).

The Word of God as the Divine Attractor in Evolutionary Emergence

How does this understanding of evolutionary emergence relate to the incarnation? Again, Rahner's thought offers inspiration for the following response.[8] Jesus in his humanity can be seen, like the rest of us, as sprung from the emergent universe, made of the elements such as carbon that have been formed in stars, a product of

---

[7] Walter Kasper, *The God of Jesus Christ* (London: SCM, 1983), 227.
[8] Rahner, *Foundations of Christian Faith*, 178–203.

biological evolution on Earth. But as a creature, Jesus of Nazareth can also be seen as the unique and unforeseeable culmination of this process in that he is the creature who gives himself radically into the mystery of God—in his life lived in radical and complete love for God and for others, which finds ultimate expression in his death on the cross. In Jesus, part of creation transcends itself in self-giving love to the one he called "Abba! Father!" and responds to divine love in a most profound "yes!"

From the perspective of his humanity, then, Jesus is the unique self-transcendence of creation to God. From the perspective of his divinity, Jesus is the unique, irreversible culmination of God's self-bestowal to a world of creatures. He is the self-giving of God, the eternal Word of God made flesh in our history, through the power of the Spirit who overshadowed Mary (Luke 1:35). Jesus is, then, both God's radical self-bestowal to creation and, in his human life and death, the radical response of creation to God. He is both God's forgiving, healing, liberating embrace of creation in love and creation's "yes!" to God.

In the death of Jesus, a part of this evolutionary world is handed over in freedom into God in complete love. In the resurrection and the ascension of the risen Christ, this creaturely reality of Jesus is fully taken into God and irrevocably adopted as God's own reality. In the divine economy of salvation, this is an event for the whole of creation. What occurs in Jesus, as part of the physical, biological, and human world, is ontologically the "beginning of the glorification and divinization of the whole of reality."[9] The future of the emergent universe is promised in the resurrection of the crucified Jesus. It is to be transfigured in the risen Christ.

Can more be said about the relationship between the evolving universe and the Word of God who is made flesh in Jesus? I find a helpful beginning in the image of the attractor proposed by the Polish philosopher and Archbishop of Lublin, Jósef Życiński (1948–2011). In his view, the traditional concept of God as Author of Cosmic Design, a God who operates from a precise preordained

---

[9] Karl Rahner, "Dogmatic Questions on Easter," *Theological Investigations* 4 (Baltimore: Helicon Press, 1966), 129.

divine plan, no longer fits well with what we know of the natural world and cannot be maintained. He insists that theology must take account of what science finds in the natural world, including its discontinuities, evolutionary bifurcations, and random processes. In the light of the contingency of such natural processes, how can we find better ways to speak of God as giving direction to evolution? He suggests replacing the analogy of the divine planner with one taken from the role of an attractor in dynamic systems. God, then, can be thought of as the "Cosmic Attractor" of evolution.[10]

Życiński points to the use of the concept of the attractor in both mathematics and in physics, particularly the study of the physics of nonlinear systems, where the system is found to be drawn to a particular state. It appears as if this state is attracting the system to itself—"The essential factor in this process is the dynamic by which the system is directed locally toward a physical state which is as yet unrealized, but nevertheless gives the appearance that it is 'attracting' to itself, at the given stage, the evolution of the system itself."[11] Although Życiński does not mention it, the image of the attractor is also found in a quite different form in astronomy, and here the attraction is due to gravity: the "Great Attractor" names a part of the universe with enormous mass that has been seen as exercising a powerful gravitational pull on the Milky Way and thousands of other galaxies.[12]

In taking up the analogy of the attractor, Życiński acknowledges its similarity to Teilhard de Chardin's image of the Omega Point drawing evolution to its fulfillment, and to Alfred North Whitehead's idea of God as the noncoercive Poet of the World. He points, as well, to the connection between his idea of God as the Attractor and the thought of systematic theologians—Karl Rahner with his

---

[10] Jósef Życiński, *God and Evolution: Fundamental Questions of Christian Evolutionism* (Washington, DC: Catholic University of America Press, 2006), 161–64.

[11] Ibid., 162.

[12] Recently it has been discovered that much of this gravitational attraction comes from a massive cluster of galaxies beyond the Great Attractor, near the Shapley Supercluster.

concept of God as "Absolute Future" and Jürgen Moltmann, Wolfhart Pannenberg, and Ted Peters, who see God as the "Power of the Future."[13]

Others have also taken up the image of the attractor. Ilia Delio speaks of God as the "strange attractor" luring creation toward the optimal good and, in another context, of Christ as the "strange attractor" drawing those beyond the churches to a spiritual search, to compassion for others and to commitment to the community of life on Earth.[14] The attractor image is closely related to an important theme in the work of John Haught who, with Rahner, sees creation as directed to the Absolute Future that is God. He understands this Power of the Future as acting not just on human beings but on the whole cosmic process, so that "the *entire universe* is always being drawn by the power of a divinely renewing future."[15] Theologically speaking, Haught says, "biological evolution is part of a great cosmic journey into the incomprehensible mystery of God."[16] He sees God as always calling life "into the freshness of the future."[17]

I will take up the image of the attractor in a particular way, focusing it in a trinitarian and fully christological theology that builds on Athanasius. My suggestion is that the eternal Word of God can be imaged as the divine Attractor in the evolutionary emergence of the universe and its individual entities, and that the Word made flesh, Jesus crucified and now risen from the dead, can be thought of as the Attractor not only of evolutionary emergence but also of the final transformation and fulfillment of the whole creation. This attraction of the Word is not any kind of physical force that could be discovered empirically, but the divine act that we call God's continuous creation and God's salvation of a world of creatures. The power of attraction

---

[13] Życiński, *God and Evolution*, 162–64.

[14] Ilia Delio, *The Humility of God: A Franciscan Perspective* (Cincinnati, OH: St. Anthony Messenger Press, 2005), 75–85; *The Emergent Christ: Exploring the Meaning of Catholic in an Evolutionary Universe* (Maryknoll, NY: Orbis Books, 2010), 142–46.

[15] John F. Haught, *God after Darwin: A Theology of Evolution*, 2nd ed. (Boulder, CO: Westview Press, 2008), 97.

[16] Haught, *Making Sense of Evolution*, 75.

[17] Ibid., 73.

is the bond of divine creative and saving love, the presence and action of the indwelling Spirit. The Holy Spirit, the power of the new, *is* this indwelling attraction, drawing all things to the Word of God.

Earlier I discussed Athanasius's view of each creature as existing by partaking of the Word through the indwelling Spirit, and noted that for him, it is the boundless generativity of the divine life of the Trinity that accounts for all the generativity involved in bringing creation to life. For him, God is a Light with its everlasting Radiance that enlightens us in the Spirit, a Fountain always pouring forth a River of living water from which creatures drink in the Spirit, a Father eternally begetting the Son in whom we participate by adoption as God's children in the Spirit. Such is the dynamic nature of divine Communion. The eternal generation of the One who is Wisdom, Word, and Radiance is the basis for the emergence of all the creatures in their specific nature. Each is a reflection of divine Wisdom, a kind of icon of Wisdom. Each partakes of the Word in the Spirit.

In evolutionary terms, this Word can be understood as the divine Attractor, drawing into existence galaxies, stars, and planets, and then, on Earth, calling into existence through evolutionary processes all the diverse species of microbes, insects, birds, fish, plants, and animals, including human beings. The divine Word draws each species to its own identity and place in evolutionary emergence. Not just each species, but each member of each species, each sparrow, is held in the divine memory and embraced in the divine love, as a word of the Word, an expression of divine Wisdom in our world.

In this view, then, the incarnation of the Word is the incarnation of the Attractor of evolutionary emergence. As John's gospel tells us, all things were made through the Word of God (John 1:3), and this Word of creation is made flesh in our midst (John 1:14). If this text is read in the light of evolution, then it can be interpreted as speaking of a profound connection between the evolutionary emergence of a universe of creatures and what happens in the birth, life, death, and resurrection of the Savior. In the theology of John's gospel, and in the subsequent Christian theology of the incarnation, the whole creation is directed toward this event and is transformed by it.

In the resurrection of the crucified Jesus, the Word of God is forever flesh, but earthly, bodily reality now transfigured in glory.

The risen Christ is the embodiment of new creation, its promise and its beginning. The incarnate Word, the crucified and risen Christ, is now the Attractor of the whole creation, not just to its evolutionary existence, but to its transfiguration and fulfillment. And the Holy Spirit is the enabling power at work in this whole process—the very attraction, the drawing power of love, the life-giving presence at work in it all.

To say that the risen Christ is the Attractor of the whole process of evolutionary emergence is to speak in evolutionary terms about the promises of a future for all things in Christ that are already contained in the Scriptures: "the creation itself will be set free from its bondage to decay and will obtain the freedom of the glory of the children of God" (Rom 8:21); "For in him all the fullness of God was pleased to dwell, and through him God was pleased to reconcile to himself all things, whether on earth or in heaven, by making peace through the blood of his cross" (Col 1:19-20); "With all wisdom and insight he has made known to us the mystery of his will, according to his good pleasure that he set forth in Christ, as a plan for the fullness of time, to gather up all things in him, things in heaven and things on earth" (Eph 1:8-10).

One of the advantages of the analogy of the Attractor is its non-anthropomorphic character. It points to the fulfillment and transfiguration of a cosmic world far beyond the human. But I see it as having the further advantage that it can also be understood in a very human and quite personal way, as offering meaning for human beings on their spiritual journeys. The gospels tell of a Jesus who attracts great crowds in Galilee. He attracts not only the adults but also the children, telling those who would push them away: 'Let the little children come to me; do not stop them; for it is to such as these that the kingdom of God belongs" (Mark 10:14). He draws followers to himself, involves them in a lifelong relationship, and calls them to become active participants in his mission on behalf of the kingdom of God. Those who are attracted to him and are caught up in the vision of God's kingdom he proclaims include people who come for healing and liberation, young and old, women and men, and particularly the poor of Galilee.

In the Wisdom books of the Bible, divine Wisdom is presented as an attractive Woman who invites all to the feast she has prepared:

"Come, eat of my bread and drink of the wine I have mixed" (Prov 9:5); "Come to me, you who desire me, and eat your fill of my fruits" (Sir 24:19). In the New Testament, Jesus-Wisdom invites all kinds of people to his inclusive table, including not only the poor but also many who are public sinners (Mark 2:15-17). In a particular way, as the Wisdom of God in our midst, Jesus reaches out to draw to himself all those who struggle in life with weariness, pain, and grief: "Come to me, all you that are weary and are carrying heavy burdens, and I will give you rest. Take my yoke upon you, and learn from me; for I am gentle and humble in heart, and you will find rest for your souls. For my yoke is easy, and my burden is light" (Matt 11:28-30).

At the most human level, Jesus-Wisdom draws us in the ups and downs of our existence, attracting us, even in our resistances, into the new, not only in our lives, but also, and above all, in our deaths. In John's gospel, we are told explicitly that it is the Father's doing when we are drawn to Jesus, and it seems that the attraction is the gift of the Spirit: "No one can come to me unless drawn by the Father who sent me; and I will raise that person up on the last day. It is written in the prophets, 'And they shall all be taught by God.' Everyone who has heard and learned from the Father comes to me" (John 6:43-46). Jesus invites all those who thirst to come to him, to receive the gift of living water, the Spirit who wells up from within: "On the last day of the festival, the great day, while Jesus was standing there, he cried out, 'Let anyone who is thirsty come to me, and let the one who believes in me drink'" (John 7:37-38).

Perhaps the deepest theology of Jesus as the divine Attractor is found in the image of Jesus being lifted up and attracting all to himself as the crucified and risen one. The context is Jesus speaking of his soul being troubled by the coming hour of his death, but praying nonetheless, "Father, glorify your name." A voice from heaven replies: "I have glorified it, and I will glorify it again" (John 12:28). Jesus then proclaims "And I, when I am lifted up from the earth, will draw all to myself" (John 12:32). This last text is in my own translation. In the NRSV translation, which I am following in this book, it is "all people" that Christ draws to himself, because the Greek adjective "all" is in the masculine form. In an important variant reading, however, it is neuter in form, in which case it would refer to "all things." Whether or not the focus of the original text was on all people or all things, biblical faith generally sees human beings as being saved along with the wider creation. The many texts that speak of "all things" being transfigured in the risen Christ (Rom 8:19-23; 1 Cor 15:28; Col 1:15-20; Eph 1:9-10; Rev 5:13-14) support the idea that it is theologically appropriate to see Jesus lifted up in the cross and resurrection as the divine Attractor not only for human beings but also the whole universe of creatures with them.

This theological claim, of course, is not warranted by the sciences but only by Christian faith in the crucified and risen Christ. Biological science, with its understanding of natural selection, cannot tell this story. It works only with the methodologies proper to the natural sciences. Methodologically, it is rightly committed to a completely naturalistic explanation. It is only when it is claimed that that these methods fully explain the whole of reality (ontological naturalism) that theology has to disagree. Theology can gratefully accept all the findings of the natural sciences working according to their own methodologies, and thus discover a great deal more about the world that it sees as God's creation. In dialogue with evolutionary science, but based on its own resources, theology can tell a story of evolutionary emergence as participation in the life of God. But this is not in any way a simplistic story of evolutionary optimism. The natural world itself offers no guarantees that we will not face an extreme catastrophe as a result of a natural phenomenon (such as a large meteorite colliding with Earth), or as a result of human actions.

At a deeper level, at the center of Christian faith, at the heart of the incarnation, we find God identified with the suffering of the world in the cross of Christ, something taken up in the next chapter. So, the Christian story is not a story of evolutionary optimism. But it is a story of the absolute promise of God, given in the resurrection of the one who was crucified. It is this that makes the theological story of evolutionary emergence a story of Christian hope in the midst of the messiness and unpredictability of natural processes and in the midst of human failure and sin. It is a story of hope for the final fulfillment of the whole emergent universe of creatures in the crucified and risen Word of God, who in the Spirit, the Energy of Love, draws the whole creation to its new, transfigured state.

Chapter 5

The Costs of Evolution
and the Divine Passion of Love

In earlier times it seemed obvious that suffering and death could be interpreted, at least in part, as the result of human sin. Of course it remains clear today that sinful human behavior is responsible for an enormous amount of suffering, death, and ecological devastation on our planet. But we now know that while life on Earth began about 3.7 billion years ago, modern humans appeared only about 200,000 years ago. Clearly, modern humans are not solely responsible for the pain and death that are part of the natural world. Evolutionary biology has taught us that evolution is an extremely costly process, and that the costs of evolution are built into the process from the beginning.

The Costs of Evolution

In recent years, biologists have discovered more about the place of cooperation in evolution.[1] But along with this cooperation they see competition, predation, pain, and death as intrinsic to evolutionary processes. In an evolutionary view, death is not the result of something gone wrong with the process. It is simply part of the process. The evolution of the eye, for example, depends upon the cycles of life and death over countless generations, in which random genetic

---

[1] See, for example, Martin A. Nowak, with Roger Highfield, *SuperCooperators: Altruism, Evolution, and Why We Need Each Other to Succeed* (New York: Free Press, 2011); Edward O. Wilson, *The Social Conquest of Earth* (New York: Liveright Publishing Corporation, 2012).

mutations give an advantage in adapting to an environment and reproducing. Natural selection needs the cycle of generations. Death is the price paid for a world with developed forms of life, including blue wrens, whales, and human beings. Death is the price paid for a world in which there are the wings of an eagle, the powerful legs of a red kangaroo, and the brain of a human being.[2]

Evolution involves what seems like wasteful abundance of some life-forms, competition for resources, and the predator-prey relationship. Pain is part of the process and has survival value, acting as a warning and a spur to action. In the history of life on Earth, there have been five major extinctions, including one that occurred 250 million years ago, when most of life was annihilated, and another 65 million years ago, when the dinosaurs and many other species disappeared. Today, of course, because of human actions, we face a new, devastating, sixth extinction, a tragic loss of Earth's biodiversity.

How are we to think about the Creator in relation to the costs of evolution? If the costs of evolution cannot be attributed to human beings, then it seems the responsibility rests with the Creator—at least for those who believe in God. An evolutionary view of the world thus intensifies the age-old theological problem of evil, both because of the sheer scale of the suffering and because it seems the Creator alone is responsible. Why does God create in a way that is so devastatingly costly for so many creatures?

In my view, there is no adequate intellectual answer to this question. But an evolutionary theology of God must address the issue as best it can.[3] It needs to be able to speak of the God of boundless compassion revealed in Jesus as the God of evolutionary emergence. I see three fundamental theological responses to the problem of evil that can be offered on the basis of the Christian gospel. These responses offer a basis for hope, but they do not resolve the problem of evil. They do not tell us why the world is the way it is. The first and most important of these responses has been discussed in earlier

---

[2] Ursula Goodenough, *The Sacred Depths of Nature* (Oxford: Oxford University Press, 1988), 143–51.

[3] See Christopher Southgate, *The Groaning of Creation: God, Evolution, and the Problem of Evil* (Louisville, KY: Westminster John Knox Press, 2008).

chapters, that God hears the groaning of creation, embraces the world of creatures in the incarnation and in the cross, and promises creation's deliverance and fulfillment in the risen Christ. The second, taken up here, is that God cares passionately about creation, is lovingly present to it, and suffers with it in its groaning. The third, discussed in the next chapter, is that God humbly respects and waits upon the unfolding of creation according to its own proper processes.

Is God Free from Suffering? Does God Suffer with Creatures?

Does God suffer with us creatures when we suffer? This always difficult question has been unavoidable in theology since the horrors of the First and Second World Wars and, above all, the Holocaust. As I have been indicating, I see it as unavoidable, as well, in an evolutionary and ecological theology of creation. Confronted by the costs of evolution, it becomes necessary to ask: Is God free from the suffering of creation and, if so, in what sense? Does God suffer with suffering creation and, if so, in what sense?

In the ancient Greek world, the word *apatheia* referred to one who was without *pathos*, without suffering or passion. It suggested a state of mind where a person is not controlled by turbulent emotions or by external forces. For the Greeks, *apatheia* often connoted something quite different from the modern English word *apathy*, which carries the highly negative sense of being coldly indifferent toward the suffering of others. The meaning of *apatheia* was, perhaps, closer to equanimity, a state of inner freedom, the goal of Buddhist practices as well as many forms of Christian spirituality. This kind of freedom can, of course, coexist with profound compassion for other creatures. In the early church, the Christian community could think of the blessed in heaven and also the martyrs facing death as possessing this gift. In the case of the martyrs, *apatheia* was understood not as freedom from physical suffering or fear, but as the freedom to face them that comes from dying in Christ.[4]

---

[4] Paul Gavrilyuk, *The Suffering of the Impassible God: The Dialectics of Patristic Thought* (Oxford, UK: Oxford University Press, 2004), 69–75.

Another word for this same quality that has its roots in Latin is *impassibility*. It too has the meaning of not being subject to suffering and not being ruled by the passions. The truly wise person could be thought of as possessing something of this quality, retaining an inner freedom in all the turbulence of life. In some philosophical circles of the Greco-Roman world, God was understood as transcendent and impassible, as opposed to mythological gods who were subject to all-too-human passions. Early Christian theologians consistently saw the biblical God as possessing this quality, but for them it was radically qualified by their conviction that the Word of God suffers in Jesus Christ, above all, in his cross.

The Paradox:
The God Who Is Free from Suffering Suffers in Christ

All theological talk of God suffering with creatures has to deal with the strong Christian tradition that has seen God as free from suffering. In recent times, some have simply dismissed the patristic tradition of divine impassibility as an import from Hellenistic thought that distorts the vitality of the biblical view of God. But this view fails to recognize that for Irenaeus, Athanasius, and many others, the concept of divine impassibility defends the biblical concept of the radical otherness of God. For the great patristic theologians, the theology of impassibility was a way of resisting all tendencies to see God as a creature trapped within the vicissitudes of creation or as at the mercy of changing emotions or as arbitrary and fickle like the gods of Greek mythology. Impassibility upholds the constancy and the *fidelity* of the biblical God in creation and salvation.

Paul Gavrilyuk challenges the scholarly construct that portrays a kind of "fall" in early Christian theology away from the biblical view of a suffering God and into a Greek view of an impassible God. He insists that the biblical canon portrays God as both participating actively in history and as divinely transcendent. And, he argues, the patristic writers advocate not an absolute impassibility that removes God from the world or denies emotions such as love to God but a *qualified* impassibility, one which rejects the attribution of unworthy creaturely emotions to God. This kind of impassibility does not completely rule out "God-befitting" emotions. Rather, it functions as "a kind an apophatic qualifier of all divine emotions and as the marker of the unmistakably divine identity."[5] Emotions such as love, compassion, and generosity can be properly attributed to God, when it is acknowledged that God's love, compassion, and generosity are of a Godlike kind, infinitely beyond all human emotions. What are ruled out by the "apophatic qualifier" of impassibility are fickleness, arbitrariness, and inconstancy, and all the emotions and passions that are unworthy of God but found in mythological gods and human tyrants, including lust, jealousy, vengeance, and violence.

The patristic defense of God's impassibility is deeply interconnected with the central Christian conviction that the incarnate Word of God truly suffers on the cross. As Irenaeus puts it, "The Impassible One became capable of suffering."[6] This often repeated theme expresses a real, enduring, Christian paradox: Because the Word is truly divine, the Word possesses divine impassibility; because this Word becomes flesh, the Word truly suffers in the cross of Christ.[7] Melito of Sardis sums up this paradox: "the impassible one suffers."[8]

Athanasius insists that the Word of God possesses impassibility because the Word possesses the divine essence, which he sees as impassible. Furthermore, it is because the Word is impassible that the Word made flesh can bring creation to its liberation from suffering and death. Through the death and resurrection of Jesus, we creatures are deified and participate in divine impassibility. Yet this occurs through the incarnate Word making the sufferings of creaturely existence the Word's very own. Athanasius explicitly embraces this as paradox: "And it was a paradox that he both suffered and did not suffer; he suffered insofar as his own body suffered and he was in this suffering; but he did not suffer, because the Logos is God by nature and is impassible."[9]

[5] Gavrilyuk, *The Suffering of the Impassible God*, 173.

[6] Irenaeus, *Adversus haereses* 3.16.6.

[7] Michael Figura, "The Suffering of God in Patristic Theology," *Communio: International Catholic Review* 30, no. 3 (2003): 366–85.

[8] Melito of Sardis, Fragment 13 from *On Soul and Body*, Sources Chrétiennes 123, 238.

[9] Athanasius, *Epistola ad Epictum* 6, Patrologia Graeca 26, 1060 C.

The suffering of the cross is not simply that of the human nature. It is the suffering of the eternal Word of God, the impassible Word. The eternal Word, the impassible Word, suffers in Jesus Christ. In his discussion of this issue, Karl Rahner does not seek to resolve this paradox, but deepens it. He holds to the unchanging impassibility of the Word of God, but he says that in the incarnation the unchangeable God *becomes* something: "God can become something, he who is unchangeable in himself can *himself* become subject to change *in something else*."[10] The Word of God changes and suffers, in the other of creaturely reality. The incarnation tells us something totally unpredictable. God who is infinite fullness of life and love can "empty" God's self in love in creation and incarnation (Phil 2:4-8). At the heart of Christian faith, Rahner says, there is "the self-emptying, the coming to be, the κένωσίσ and γένεσισ of God himself, who can come to be by becoming another thing" without changing in God's own proper reality. The incarnate Word of God changes and suffers in the incarnation but, paradoxically, still possesses the fullness of divinity.

In my view, this paradox cannot be resolved. It remains necessary if we are to remain faithful to two biblical/theological insights: 1. That in the life and death of Jesus Christ the eternal Word of God truly suffers, since the suffering truly belongs to the person of the Word and not solely to the human nature; 2. That the Word is and remains truly God, and therefore remains always in the undivided, glorious Communion-in-Love that is the holy Trinity. This divine Communion-in-Love is the source of existence for all creatures. And participation in this Communion of the Trinity will be the healing, liberation, and final fulfillment of the whole creation.

I think that Jürgen Moltmann and Hans Urs von Balthasar, in their quite different ways, are in danger of appearing to undermine this Communion as fullness in love when they push the economic suffering and the self-emptying of the Word back into the eternal trinitarian relations, apparently making suffering and self-emptying essential to trintarian life. Moltmann takes the cry of abandonment

---

[10] Karl Rahner, "On the Theology of the Incarnation," *Theological Investigations* 4 (Baltimore, Helicon Press, 1966), 113.

on the cross as the expression of a *real abandonment* that reaches into the heart of the Trinity: "On the cross the Father and the Son are so deeply separated that their relationship breaks off." He speaks of the "breakdown of the relationship that constitutes the very life of the Trinity. . . . The love that binds the one to the other is transformed into a dividing curse."[11] What happens in the cross "reaches into the innermost depths of the Godhead, putting its impress on the trinitarian life in eternity."[12]

Balthasar projects the self-emptying (kenosis) at work in Christ and his cross back into the eternal Trinity. He sees the begetting of the Word by the Father as the original kenosis which is the basis for the self-emptying kenosis of creation and of the cross.[13] The Father "strips himself, without remainder of his Godhead and hands it over to the Son."[14] The generation of the Son brings about an "absolute, infinite distance" and "difference" within the Trinity, which the Holy Spirit maintains and also bridges. The suffering and death of the cross are understood as embraced within the "fire" of the relations of the immanent Trinity. Balthasar can speak of the "fire of love" as what each divine person is for the other, and of the "fire of suffering" as the "essential feature" of the triune God.[15]

I am not convinced that we should apply the suffering of the cross (Moltmann), or the word kenosis with its connotation of the extreme suffering of the cross (Balthasar), to the inner life of the divine persons. This is a speculative strategy, which risks undermining the unity of the divine persons, the beatitude and glory of divine life, and the fullness of communion that is the promised

---

[11] Jürgen Moltmann, *The Trinity and the Kingdom of God: The Doctrine of God* (San Francisco: Harper & Row, 1981), 80.

[12] Ibid., 81. While I distance myself from some of Moltmann's positions, I think contemporary theology owes him a great deal for bringing the issue of the suffering of God to the center of theological attention, for provoking a vitally important discussion, and for eliciting responses from others.

[13] Hans Urs von Balthasar, *Theo-Drama: Theological Dramatic Theory*, vol. 4: *The Action* (San Francisco: Ignatius Press, 1994), 319–28; *Theo-Drama: Theological Dramatic Theory*, vol. 5: *The Last Act* (San Francisco: Ignatius Press, 1998), 212–16.

[14] Balthasar, *Theo-Drama*, vol. 4: *The Action*, 323.

[15] Balthasar, *Theo-Drama*, vol. 5: *The Last Act*, 268.

future of creation in its groaning. I find myself agreeing with the critical remarks of Gilles Emery. He says of Moltmann's view that in attempting to explain suffering "he does so through divinizing suffering and rendering it eternal, as an event that is constitutive of the divine persons."[16] Of Balthasar's idea, he says that "it leads one to understand the unity of the three persons in terms of a dialectical event of distance."[17]

In our suffering as creatures, I do not think we are helped at all by the idea of a God of inner-trinitarian suffering or distance. What helps is a God who is with us, who feels with us, out of divine compassionate love. I think an analogy from family life can be helpful here. A daughter in terrible trouble is not helped by a painful distance or division between her parents, but by them being united in love for her, by them both being with her, suffering with her in her trauma, and doing every possible thing they can for her, in what can be costly, self-emptying, suffering love. There is nothing consoling or freeing for a daughter in her parents becoming alienated from one another.

In something of a similar way, divine love for suffering creation is grounded, I believe, not in divisions or distance between the divine Persons but in the fullness of the Communion of the Trinity. Bruce Marshall is right when he insists that in the cross the eternal Word and the Father love each other perfectly in the unity of the Holy Spirit: "Just because the Father and the Son abide with one another even there, the Son's passion is our salvation. . . . His passion is our way into the love which infinitely and unalterably goes with being God, into the perfect blessedness and rest of God's life."[18]

Is there another way to think about God suffering with suffering creation? Is there an alternative to those, like Moltmann and

[16] Gilles Emery, "The Immutability of the God of Love and the Problem of Language Concerning the 'Suffering of God,'" in *Divine Impassibility and the Mystery of Human Suffering*, ed. James F. Keating and Thomas Joseph White (Grand Rapids, MI: Eerdmans, 2000), 44.

[17] Emery, "The Immutability of the God of Love," 52.

[18] Bruce D. Marshall, "The Dereliction of Christ and the Impassibility of God," in *Divine Impassibility and the Mystery of Human Suffering*, 298.

Balthasar, who locate the suffering and self-emptying of the cross within the trinitarian relations, and to those, like Thomas Weinandy and Brian Davies, who oppose the idea of divine suffering with creatures?[19] I find a helpful resource for responding to this question in the brilliant and influential patristic thinker, Origen of Alexandria (185–254).

The Passion of Love

Origen wrote eight volumes in reply to the pagan writer Celsus, who had argued that the incarnation is impossible because it is simply unthinkable for God to undergo a change from a better to a worse state. In his response, Origen holds fast to the paradox that while God remains unchanging in the divine essence, yet God descends and becomes deeply involved with creation for the sake of providing for creatures in the economy of salvation.[20] He says that, in the incarnation, the Word comes to the human world like a doctor to heal the illness of sin, but remains unchanged in the divine nature. Origen is well aware of biblical passages that speak of God having human-like passions and emotions, but thinks that these should not be taken literally: "We do not attribute human passions to God."[21]

But Origen clearly does think that it is appropriate to speak about God suffering for and with creation in the specific sense that God suffers out of the eternal divine passion of love for creatures. In a particularly beautiful passage from his *Commentary on Ezekiel* he writes:

> Let me offer a human example; then, if the Holy Spirit grants it, I will move on to Jesus Christ and God the Father. When I speak to a man and plead with him for some matter, that he should have pity on me, if he is a man without pity, he does not suffer anything from the things I say. But if he is a man of

---

[19] Thomas Weinandy, *Does God Suffer?* (Edinburgh: T & T Clark, 2000); Brian Davies, *The Reality of God and the Problem of Evil* (London: Continuum, 2006).
[20] *Contra Celsum* 4.14.
[21] *Contra Celsum* 4.72.

gentle spirit, and not hardened and rigid in his heart, he hears me and has pity upon me. And his feelings are softened by my requests. Understand something of this kind with regard to the Savior. He came down to earth out of compassion for the human race. Having experienced our sufferings even before he suffered on the cross, he condescended to assume our flesh. For if he had not suffered, he would not have come to live on the level of human life. First, he suffered; then he came down and was seen (cf. 1 Tim 3:16). What is this suffering that he suffered for us? It is *the passion of love* (*caritatis est passio*). The Father, too, himself, the God of the universe, "patient and abounding in mercy" (Ps 103:8) and compassionate, does he not in some way suffer? Or do you not know that when he directs human affairs he suffers human suffering? For "the Lord your God bore your ways, as a man bears his son" (Deut 1:31). Therefore God bears our ways, just as the Son of God bears our suffering. The Father himself is not without suffering. When he is prayed to, he has pity and compassion; he suffers *the passion of love* (*patitur aliquid caritatis*) and comes into those in whom he cannot be, in view of the greatness of his nature, and on account of us he endures human sufferings.[22]

Henri De Lubac highlights this text in his discussion of Origen's view of God, exclaiming that it is "one of the finest pages, without doubt, one of the most humane and the most Christian pages we have from him . . . An astonishing, wonderful text!"[23] He points out that this theological text is all the more significant because Origen clearly does accept the idea of divine impassibility and is choosing his words carefully. De Lubac sees Origen as striving in this passage to express the counterintuitive depth of divine mercy revealed

---

[22] Origen, "Homilies on Ezekiel," 6.3 in *Homilies 1–14 On Ezekiel*, trans. Thomas P. Scheck, Ancient Christian Writers 62 (New York: Newman Press, 2010), 92–93. I have modified the translation of the two italicized phrases, highlighting Origen's declaration that the Word suffers "the passion of love," and his use of an equivalent expression of the Father.

[23] Henri De Lubac, *History and Spirit: The Understanding of Scripture according to Origen* (San Francisco: Ignatius Press, 2007), 275.

in Christ that overturns all rational expectations: "The revelation of Love overturns all that the world had conceived of divinity."[24]

At the beginning of this passage Origen states his intention—he will begin from a human example of compassion and, on this basis, consider first Jesus Christ and then God the Father. For his human example he imagines a situation where he finds himself in extreme need and asking for help and mercy from another. He describes two very different responses he might get. The first is from a man with no pity—such a man, Origen tells us, does not *suffer* anything when hearing the sad story. The second response is very different, from a man of gentle spirit, not hard and rigid, who hears the need of the other and has pity. What is noteworthy is Origen's comment that the man without pity does not "suffer" anything when hearing the sad story. This person, Origen seems to be thinking, does not feel the pain of another, and therefore is not moved by it. He does not "walk in the shoes" of the other. In the second case, we find a man capable of empathy, who can imagine himself in another's shoes, who has not hardened himself against feeling for the other.

Origen then applies this line of thought to the Savior. The man capable of suffering with another, he suggests, gives us some insight into the Word of God. The Word comes to us, he says, out of compassion. The Word experiences and feels with us in our sufferings *before* the incarnation. If the Word had not suffered with us, and did not feel our pain in some way, then the Word would not have come to us. In his view, the motivation for the incarnation is the divine feeling, the divine passion for creatures in their need. The Word comes to us, Origen tells us, because of the divine *"passion of love."*

What of the Father, the Creator of the universe? Origen is emphatic that the Father is not without suffering, and not impassible in this sense, but like the Word suffers with creatures. Earlier, I made it clear there is an important sense in which Origen thinks of God as being impassible. But in this text, he shows that there is a real (analogous) sense in which the Father does suffer with us out of divine compassion—in the divine passion of love. He says that

[24] Ibid., 277.

when we approach God with our need in prayer, the Father is moved with com-passion, and suffers with us. The man who "suffered" in hearing the story of another in need is a pale but apt image of the compassionate God. God the Father, Origen says, suffers *the passion of love.*

Of course, at the heart of the Christian tradition, we have Jesus' own image of this divine compassion in his wonderfully vivid picture of the father in his parable of the Prodigal Son: "But while he was still far off, his father saw him and was filled with compassion; he ran and put his arms around him and kissed him" (Luke 15:20). Not only this text but also all the liberating words and healing actions of Jesus witness to this same divine compassion. This same divine attribute was long expressed in the faith of Israel, at times in the idea of God's womb-like mercy and the image of God as Mother: "As a mother comforts her child, so will I comfort you" (Isa 66:13). With Origen, then, I think it is appropriate to speak, by way of analogy, of God's passion of love, and of God who suffers with creation in the sense that God feels with human beings in their need and for the whole creation in its groaning.

Theologically, this passion of love can be seen as grounded in the eternal, mutual love of the divine persons in the life of the Trinity. It is the eternal love of the divine Communion that embraces creatures, through the Word and in the Spirit. This dynamic, transforming, inclusive love is identical with the divine nature—"God is love" (1 John 4:8). The divine passion of love for creatures is better understood, I believe, not as springing from division or distance between the trinitarian persons, as in the positions of Moltmann and Balthasar, but rather as grounded in the mutual giving and receiving of love, the abiding in one another, the perichoresis, of the undivided Communion of the Trinity.

What in this context, then, is the meaning of the tradition of divine impassibility? What does it rule out in God? What Origen would rule out, I suspect, is: 1. That suffering is imposed on God from outside; 2. That the divine passion of love is corporeal; 3. That God suffers in the way humans do because of their own sin; 4. That God suffers change because God is ruled by the unruly, passing emotional states that can dominate human beings; 5. That

God changes in God's constancy, fidelity, will to save, or capacity
to bring creation to its fulfillment.

As Elizabeth Johnson says, analogical speech about the suffering
of God does not intend to suggest that God suffers by necessity
or passively, but points, rather, "to an act of freedom, the freedom
of love deliberately and generously shared."[25] The words "suffer
with," and "passion of love" are used of God strictly analogically,
which means that the infinite difference between human experience
and divine love is fully recognized, and these qualities are affirmed
of God in a way that transcends all human experience of them.
The divine passion of love is infinitely beyond human capacities
for empathy with others.

Walter Kasper points out that this kind of divine vulnerability
in love is not the expression of a lack in God, but the expression of
divine fullness. This kind of vulnerability expresses the capacity
to love in a transcendent and divine way. God does not suffer from
lack of being, but suffers out of love which is the overflow of the
divine being. This kind of suffering does not befall God, then, but
expresses the divine freedom to love. The self-giving love involves
allowing the other to affect oneself. This is not a passive "being-
affected," but a free, active allowing the other to affect oneself. Because God is love, God can suffer with us.

Kasper sees the suffering of the cross as the true revelation of God.
He points out that the eternal distinction between the Father and the
Son is the transcendent condition for the possibility of God's self-
giving in the incarnation and the cross. This means that from eternity
there is a place in God for the human, and a place also for a genuine
"sym-pathy" with the suffering of creatures. The God of Jesus Christ
is not simply a God of "a-patheia" but, in the real sense of the term,
a God of "sym-pathy," a God who suffers with suffering creation.[26]

Christian believers who ponder the cross of Jesus are surely right
to see in the cross the symbol of a God who loves creatures with a
compassion, a suffering with us, beyond any human capacity for

---

[25] Elizabeth A Johnson, *She Who Is: The Mystery of God in Feminist Theological
Discourse* (New York: Crossroad, 1992), 266.

[26] Walter Kasper, *The God of Jesus Christ* (London: SCM, 1983), 196–97.

being with others in their pain.[27] When the Word of God suffers in Christ, this does not mean that the human physical and emotional states of the humanity of the Savior are simply transferred to the eternal Trinity. But it does mean, surely, that the passionate love of God-with-us expressed in the cross represents the truth of the transcendent God's capacity to be with creatures in boundless generosity and passionate love. The Gospel tradition of the glorious risen Christ still bearing the wounds of the cross (John 20:27) surely suggests that the sufferings of the world are taken up into the healing, liberating, compassionate love of the triune God.

The Trinity Enfolding the Groaning Creation

Paul tells us that the whole creation is groaning, waiting to be liberated from its bondage to decay and to share in the glorious freedom of the children of God (Rom 8:22). He speaks of human beings who already have the Spirit as groaning, as they await their adoption and bodily transformation (Rom 8:23). He then goes deeper, describing the Holy Spirit as groaning with us, expressing the longing that is too deep for words (Rom 8:26).

What I have been proposing in this chapter is that we can think of the whole evolving universe, and of Earth with all of its diverse species, and of each individual creature, and of the "groaning" of each creature, as held within the passion of love of the Holy Trinity. Each species, each creature, by the very fact of creation, already participates immediately in God through the creative Word, in the Spirit—"For your immortal Spirit is in all things" (Wis 12:1). The triune God, the absolute fullness of dynamic, generative life and of love, freely chooses to create a world of creatures, and out of passionate love freely chooses to embrace these creatures anew in the unpredictable self-giving of the incarnation of the Word.

---

[27] As is evident from this chapter, while I embrace a great deal of Aquinas's theology of creation, I do not follow him in his view that the relation between God and the world is real (*relatio realis*) from the side of creatures, but one of reason (*relatio rationis*) from the side of God (*Summa Theologiae* 1.13.7). For a recent defense of Aquinas's position, however, see Emery, "The Immutability of the God of Love," 70–72.

The particularity of the incarnation teaches us that this divine passionate love is not only general but also engages with particular creatures. But this was already evident to the psalmist who sang of the mountains, the valley and plains, the springs of water, the cedars of Lebanon, and all the animals, the wild asses, the domestic cattle, the birds singing among the branches, the stork in its fir tree, the wild goats of the high mountains, the badgers in their rocks, the hungry young lions:

> These all look to you
> to give them their food in due season;
> when you give to them, they gather it up;
> when you open your hand, they are filled with good things.
> When you hide your face, they are dismayed;
> when you take away their breath, they die
> and return to their dust.
> When you send forth your spirit, they are created;
> and you renew the face of the ground. (Ps 104:27-30)

Jesus stands within this biblical tradition and takes it up in his teaching about trust in divine providence:

> Consider the ravens: they neither sow nor reap, they have neither storehouse nor barn, and yet God feeds them. . . . Consider the lilies, how they grow: they neither toil nor spin; yet I tell you, even Solomon in all his glory was not clothed like one of these. But if God so clothes the grass of the field, which is alive today and tomorrow is thrown into the oven, how much more will he clothe you—you of little faith! (Luke 12:24-28)

He points to sparrows, apparently the food of the poor and sold two for a very small coin, saying, "Not one of them will fall to the ground apart from your Father" (Matt 10:29). In the parallel in Luke we find "Not one of them is forgotten in God's sight" (Luke 12:6).

God is involved with the life and death of each creature, holding every single sparrow in the divine creative and life-giving memory. Jesus sees God as *"Abba,* Father" (Mark 14:36), as a God for human beings but also for all creatures. He looks on ravens, wildflowers, and sparrows with loving eyes, sees them as loved by God and revelatory of God. Jesus' attitude toward the creatures he finds around him seems steeped in the creation theology of the Psalms.[28]

This faith in God's loving care for each creature finds expression not only in the words of Jesus but also in those of his near contemporary, the author of The Wisdom of Solomon:

> For you love all things that exist, and detest none of the things that you have made, for you would not have made anything if you had hated it. How would anything have endured if you had not willed it? Or how would anything not called forth by you have been preserved? You spare all things, for they are yours, O Lord, you who love the living. (Wis 11:24-26)

The idea of a God who loves the living and suffers with a divine passion for creatures has practical outcomes. It leads us to make our own personal response to the divine passion of love that is directed to each of us. And it leads to a deepening of our feeling for the community of life that becomes that basis for a Christian ecological practice. As Elizabeth Johnson says, speaking of a God who is in solidarity with those who suffer can bring not only consolation but also, importantly, energy for the healing of suffering: "Knowing that we are not abandoned makes all the difference."[29] God found on the side of those who suffer becomes a powerful stimulus to act to overcome the personal and political causes of suffering. This is true of human suffering and it applies to the suffering of the rest of creation, above all, insofar as it springs from human actions.

If God is a God of passionate love for creation, then I think it can be said that our human experiences of compassion for other creatures, of genuine empathy with them, of longing for their healing and liberation, can give us a partial glimpse of the infinite depths of divine compassion for creatures: "The LORD is gracious and merciful, slow to anger and abounding in steadfast love. The LORD is good to all, and his compassion is over all that he has made" (Ps 145:8-9). The divine passion of love for creatures is constant and unchanging yet always particular, always local, and ever new.

---

[28] See Richard Bauckham, *Bible and Ecology: Rediscovering the Community of Creation* (London: Darton, Longman & Todd, 2010).

[29] Johnson, *She Who Is*, 266–67.

Chapter 6

A Humble God and the Proper Autonomy of Evolutionary Processes

In the opening chapter of 1 Corinthians, Paul confronts a community divided by competing claims for status with the cross of Jesus. He declares that Christ crucified is "the power of God and the wisdom of God" (1 Cor 1:24). Clearly, this concept of divine power is radically different from ordinary notions of power. Paul speaks of the "weakness" of God, declaring that "God's weakness is stronger than human strength" (1 Cor 1:25). What might this "weakness" of God mean? And how does it relate to the idea of God's power in continuous creation, above all, when we think of the Creator in relation to the evolutionary processes described by the sciences? I will begin to explore these questions with some reflections on divine defenselessness before taking up the related theme of the humble God. In the last section I will focus on divine humility in relationship to the integrity and autonomy of evolutionary processes.

The Defenselessness of Jesus and the Spirit

Christians confess in the Creed that God is "almighty." They understand God as the one whose immense power not only creates and sustains the whole universe but also has the capacity to bring it to its fulfillment. This conviction of God as all-powerful is fundamental to Christian faith. But the issue that is often left unclarified is the nature of this power. It is all too easy to assume that the power of God is something like that of an absolute human monarch—someone who is free to do anything, no matter how arbitrary or irresponsible. But a Christian theology cannot embrace this kind of assumption.

104

The central source for a Christian theology of divine power can be found only in the life and teaching of Jesus and its culmination in his death and resurrection. The cross reveals God's power as working through self-emptying, limitless love. The resurrection proclaims that this love is not impotent, but the most powerful thing in the universe, bringing life in its fullness to the whole creation.

The power of the cross is a power-in-love. For Christians, the totally vulnerable human being on the cross is the revelation of God. It is hard to imagine a more extreme picture of defenseless love than that of a tortured, naked human being, pinned to a cross and left to die. The point Paul seems to be making at the beginning of 1 Corinthians is that the cross of Jesus contradicts all expectations about the nature of divine power. The power of God is revealed in extreme vulnerability and self-giving love. The resurrection shows that this power-in-love is not futile, but brings forgiveness, healing, and transfiguration for human beings and for the whole creation. God is all-powerful precisely as love. To believe in God as all-powerful is to believe in the omnipotence of divine love and in this love's eschatological victory over sin, violence, brokenness, and death.

The cross manifests the vulnerability of divine love, while resurrection points to the power of this very love to heal and save. In the vulnerability of the cross there is no loss of divinity, no absence of divinity, but the revealing of the truth of divine love and divine power. The vulnerable self-giving love of Jesus in his life, death, and resurrection gives expression in our finite, creaturely world to the nature of divine omnipotence. When Jesus' cross is understood in relationship to the love poured out in his life and ministry on the one hand, and to his resurrection on the other, then the vulnerable figure on the cross can be understood as the true icon of divine power.

In his later work, Edward Schillebeeckx reflects on the defenselessness and vulnerability of God. He outlines divine vulnerability at three different levels: 1. God's defenselessness in creating a finite world; 2. God's defenselessness in Jesus Christ; and 3. God's defenselessness in the gift of the Holy Spirit.[1] Schillebeeckx chooses

---

[1] Edward Schillebeeckx, *For the Sake of the Gospel* (New York: Crossroad, 1990), 88–102.

to speak of the defenselessness of God rather than of divine power-lessness, because powerlessness and power contradict one another, whereas defenselessness and power need not. He says: "We know from experience that those who make themselves vulnerable can sometimes disarm evil!"[2] God was by no means powerless when Jesus was hung on the cross, Schillebeeckx says, but God was "de-fenseless and vulnerable as Jesus was vulnerable."[3]

As there is defenselessness in the life and death of Jesus, so there is also defenselessness of the Holy Spirit, in the church and in the wider human community. In the church, the Spirit who inspires and empowers each member is defenseless against human rejection and hardness of heart. The Spirit is surely grieved (Eph 4:30) when Christians cling to power and status, when they fail to discern the signs of the times in the light of the Word, when they hold on to what is safe and refuse to be open to the newness of God. The Spirit draws the whole human community to justice, peace, love, and care for God's creation. But the Spirit is present in the world as defense-less love, a love that is not overpowering but depends on human participation. The Holy Spirit is defenseless in the face of human rejection—but not defeated—and continues to call each person to participate in the common good of the community of life on Earth.

As God is vulnerable in the life and death of Jesus Christ and in the work of the Spirit, Schillebeeckx says, so God is vulnerable in the act of creating a world of creatures. He speaks of a kind of divine yielding in God's act of creating, when God makes room for the other. He points out that when God creates humans and chooses them as covenant partners, this partnership involves freedom and initiative on both sides. In giving creative space to human beings, Schillebeeckx sees God as willingly becoming vulnerable to human failure and sin. He says of God's act of creation that it is "an adven-ture, full of risks." The creation of human beings is a "blank cheque which God alone guarantees." By creating human beings with their finite, free wills, Schillebeeckx declares, God freely renounces the

---

[2] Edward Schillebeeckx, *Church: The Human Story of God* (New York Crossroad, 1990), 90.
[3] Ibid., 128.

power to control. This makes God "to a high degree 'dependent' on human beings and thus vulnerable."[4]

The Humble God

The God described by Schillebeeckx is not an overpowering or coercive deity but defenseless and humble. I see this view as related in some ways to Athanasius's theological vision. Athanasius sees God, in both creation and incarnation, as overcoming the radical difference, the ontological gap, between Creator and creatures by divine "condescension." As I pointed out earlier, this use of the word condescension does not have the contemporary connotation of patronizing behavior. It refers, rather, to the God who, out of sheer generosity, lovingly bends down to be intimately close to creatures, to be with them. Athanasius sees this gracious, descending love of creation and incarnation as involving the self-humbling of God in the eternal Word.

His thought is based on the christological hymn of Philippians 2:4-8, which says of the self-emptying Christ: "And being found in human form, he humbled himself and became obedient to the point of death—even death on a cross." While some of Athanasius's opponents saw the Word who comes to us in Jesus Christ as a creature who advances in divinity and glory through the cross and resurrection, Athanasius insists that the Word is from eternity truly and fully God. The Word thus has no need for self-advancing or self-promoting, but is instead self-emptying and self-humbling.[5] Christ is the descending, self-humbling God. This self-humbling is for the sake of our advancement and deification.

Of course, in the biblical and Christian tradition, humility is often seen as the stance of human beings before God. But this same tradition can also speak of the humility of the Savior who comes to us from God and reveals God to us. The long hoped-for bearer of salvation is pictured as a king who will come to us as "humble

---

[4] Ibid., 90.

[5] Athanasius of Alexandria, *Orations against the Arians* (= *C. Ar.*) 1.40, in Khaled Anatolios, *Athanasius* (New York: Routledge, 2004), 97.

and riding on a donkey" (Zech 9:9). And in the life and ministry of Jesus, and, above all, in his death, God is revealed in the humble and humiliated one. Jesus, the Wisdom of God, draws all people to himself precisely in his gentleness and humility:

> Come to me, all you that are weary and are carrying heavy burdens, and I will give you rest. Take my yoke upon you, and learn from me; for I am gentle and humble in heart, and you will find rest for your souls. For my yoke is easy, and my burden is light. (Matt 11:28-29)

Jesus is the Attractor of those who are heavily burdened precisely as the humble one. In the Gospel of John, Jesus washes his disciples' feet, a deliberate and powerful symbol of humility, and tells his disciples: "You also ought to wash one another's feet" (John 13:14). In Philippians, Paul urges the Christian community to have "the same mind" as Christ Jesus who emptied himself and humbled himself (Phil 2:5-11). If Jesus is the self-revelation of God in our world, then God is revealed in humility.

The English adjective "humble" translates the Latin word *humilis*. It is derived from the word *humus*, which refers to earth, soil, or the ground. In its origin, then, the word "humble" can mean "from the earth," "down to earth," or "grounded." In speaking of God as humble, I am referring to God's capacity to overcome the otherness between Creator and creature, to meet us human beings where we are, and to be with the whole creation where it is—in process. God's transcendent otherness is of such a kind that it enables the unthinkable nearness of a grounded, down-to-earth God, a God not only of far distant quasars but also of this handful of topsoil with its millions of microbes.

John Haught and Ilia Delio are among scholars who, in their own distinct ways, have begun to explore the connection between God's humility and the divine act of creation that takes place through evolutionary processes.[6] As I noted above, my approach is to build

---

[6] John F. Haught, *God after Darwin: A Theology of Evolution*, 2nd ed. (Boulder, CO: Westview Press, 2008), 51–60; Ilia Delio, *The Humility of God: A Franciscan Perspective* (Cincinnati, OH: St. Anthony Messenger Press, 2005), 69–88.

on Athanasius's idea of the self-humbling of God in the trinitarian action of creation and incarnation. But, of course, no one personifies the humility of God revealed in Jesus more effectively and more attractively than Francis of Assisi. Delio, a Franciscan sister, points to the way Francis not only embodies this humility in his life and actions but also speaks explicitly in his "Letter to the Entire Order" of the humble God. Francis sees the God who is present in all creation, the God who gives God's self to us in the infant born at Bethlehem, the God of the cross of Calvary, as now humbly present to us in the sublime mystery of the Eucharist, hidden in the eucharistic bread:

O wonderful loftiness and stupendous dignity!
O sublime humility!
O humble sublimity!
The Lord of the universe
God and the Son of God,
so humbles Himself
that for our salvation
He hides Himself
under an ordinary piece of bread!
Brothers, look at the humility of God,
And pour out your hearts before Him![7]

For Francis, the humble God is revealed in the Word made flesh but also in the flesh of the leper he meets and kisses. It is this humble God that Francis meets in the Eucharist. Delio also points to the theme of divine humility in the work of the great Franciscan theologian, Bonaventure. In his Sermon on the Nativity of the Lord, Bonaventure says "the eternal God has humbly bent down and lifted the dust of our nature" into unity with God. God bends low, Delio comments, so that God can meet us, finite and fragile human beings, along with other living creatures, where we are.[8]

---

[7] Delio, *The Humility of God*, 29. The translation is from *Francis of Assisi: Early Documents*, ed. Regis J. Armstrong, J. A. Wayne Hellmann, and William J. Short, vol. 1 (New York: New City Press, 1999), 118.

[8] Delio, *The Humility of God*, 51.

In Jesus—above all, in his cross—self-giving, humble love is revealed as the way of God, the way divine omnipotence works. Since this is the way that God has revealed God's self to us, I think we can only conclude that the same power-in-love that characterizes the incarnation and its culmination in the cross and resurrection can be thought of as also characterizing the divine act of creation. If this is so, then God can be understood as creating in a way that respects the limits and integrity of creaturely processes and the freedom of human beings. God can be thought of as a Creator who waits upon the proper unfolding of these processes and upon human freedom.

 ̄ The image of waiting-upon-another is not meant to convey passivity or doing nothing, but the creative and powerful waiting on another of a loving parent with a child, or a lover with the beloved, an active, nurturing, engaged love that enables the other to flourish in all his or her freedom and integrity. The love of the divine nature is such that God works with creaturely limits and waits upon them with infinite patience. By creating in love, God freely accepts the limitations of working with finite creatures.

God, the one he called Abba/Father, was with Jesus in his cross, holding him in love, and acting powerfully in the Spirit, transforming failure and death into the source of healing for the world, raising Jesus up as the beginning of life for the whole creation. Based on the nature of this divine love revealed in Christ's death and resurrection, it can be understood that in the act of continuous creation too God's love is of a kind that respects and works with the limits of creaturely processes. Divine power is the transcendent power-in-love that has an unimaginable capacity to respect the autonomy and independence of creatures, to work with them patiently, and to bring them to their fulfillment.

The Proper Autonomy and Integrity of Evolutionary Processes

I have been proposing that we might think of God as creating the evolving universe and its living creatures through the immediate presence of the Word of God, the Attractor, and the indwelling Holy Spirit, the Energy of Love. The presence of God through the

Word and in the Spirit gives rise to the whole empirical world, but this presence, by definition, cannot be discerned by the empirical methods of the natural sciences. This presence is not one that over-turns laws of nature, or intervenes to disrupt natural processes, but one that works creatively in and through these laws and processes. It is the presence of a humble God, intent upon enabling creatures to flourish on their own terms.

However, I do not follow Moltmann when he speaks of God's "withdrawing" to make space for creatures or when he writes of God's "absence" from creation.[9] I find these negative spatial meta-phors unhelpful because they obscure what I take to be the central conviction of Christian creation theology: creatures exist only be-cause God is immediately present to them though the Word and in the Spirit. As I have said repeatedly—following and building on Athanasius—entities exist and evolve only because they par-ticipate immediately in God through the eternal Word and in the Holy Spirit.

God is radically interior to every aspect of the universe from its origin and at every moment of its ongoing existence. Through the Word and in the Spirit, God is closer to each creature than the creature is to itself. In the relationship of continuous creation, God is intimately present to every aspect of the universe, from the Whirlpool Galaxy to the still unnamed insects of the Amazon rain forest. What I am proposing, then, is a strong view of God's pres-ence to each entity, but one that recognizes the humility of the God present through the Word and in the Spirit. The humble Creator respects the integrity of each creature, and of created processes, and enables creaturely autonomy to flourish.

The relationship by which God enables creation to exist and to evolve is radically unlike any other relationship between creatures. On the one hand, the created world is dependent on God for its existence and for its capacity for emergence. On the other hand, God establishes creatures in genuine difference from God's self in

---

[9] Jürgen Moltmann, *God in Creation* (San Francisco: Harper & Row, 1985), 88. For more sympathetic views of Moltmann on this point, see Haught, *God after Darwin*, 53–55; Delio, *The Humility of God*, 78–79.

their own reality. Because of the humble God's love and respect for creatures, the creature has its own otherness, integrity, and proper autonomy. The relationship of continuous creation enables true creaturely freedom.

This is expressed in a central axiom of Karl Rahner's theology. Rahner says that dependence on God and the genuine autonomy of the creature are directly, not inversely, related.[10] Closeness to God does not make creatures less autonomous. It sets them free to be themselves. In the relationship of creation, closeness to God and creaturely freedom and autonomy exist in *direct* relationship to one another. The relationship of creation creates the other as other, constantly maintaining its existence and capacity to act, while setting it free in its own autonomy. Rahner supports this line of thought by appealing to what we human beings experience of God in grace. The more closely we are drawn into the love of God, the more free we are. In the experience of grace, Rahner suggests, we find that "radical dependence grounds autonomy."[11] Creaturely integrity is not diminished because of God's indwelling creative presence. Rather, it flourishes.

What is being proposed here is that creation can be seen as an act of risk-taking love by which the ever-present and always empowering Creator enables the universe to run itself and make itself by its own laws. The Christian tradition holds that God is not passive in this process but provident and purposeful, and capable of bringing the whole work of creation to its fulfillment. God is dynamically, creatively, and lovingly involved. But God does this in a way that respects the independence and integrity of entities and relationships, while enabling, supporting, and empowering these entities and relationships.

For the Christian tradition, this risk-taking God is the God given to the world in the vulnerability of Jesus, in his life and ministry, and in his death on the cross. And this risk-taking God is also powerful and faithful, the God who brought Jesus from death to resurrection life, as the promise of fulfillment for all things. If God's power

---

[10] See, for example, Karl Rahner, *Foundations of Christian Faith: An Introduction to the Idea of Christianity* (New York: Seabury Press, 1978), 78–79.

[11] Ibid., 79.

is revealed in Christ as a power-in-love, then the divine power at work in the ongoing creation of all things can be understood as a power-in-love.

If divine love involves divine respect for created processes as well as human freedom, it may well be appropriate to say that God is not *absolutely* unlimited in achieving the divine purposes. Not only in the incarnation but also in creating, God appears to freely embrace limits in love and respect for finite creatures and for creaturely processes. God's freedom is not to be thought of on the analogy of an absolute monarch who can do absolutely anything, no matter how arbitrary. God acts according to the divine nature, and God's nature is love, a love that God works with creaturely limits. By creating in love, God embraces the limitations of creaturely processes. By creating in such a way that creatures participate in the process of becoming by evolutionary emergence, God can be understood as committed to respecting these processes.

When we look to the death and resurrection of Jesus we can find out a great deal about God's way of acting. Even when Jesus cries out "Abba, Father, for you all things are possible; remove this cup from me; yet, not what I want, but what you want" (Mark 14:36), we find that the way of this Abba is not that of overturning human freedom or the laws of nature. This Abba is with Jesus in his suffering, holding him in love, and acting powerfully in the Spirit, transforming failure and death into healing and liberation, raising him up as the beginning of life for the whole creation. God's way is revealed as that of accompaniment in love, as transformation in the Spirit, and as resurrection life.

God creates through a range of creaturely processes that involve chance and lawfulness working together. Randomness is built into evolutionary processes, particularly in the random nature of the genetic mutations that are the basis of natural selection. As Arthur Peacocke has shown, there is no necessary conflict between the idea of God achieving the divine purposes and God creating through processes that involve both randomness and lawfulness.[12] Randomness

---

[12] See Arthur Peacocke, *Theology for a Scientific Age: Being and Becoming— Natural, Divine, and Human* (Minneapolis, MN: Fortress Press, 1993), 115–21.

offers the possibility of creating a richer environment than would otherwise be possible.[13] Randomness occurs within the context of an evolving universe that has particular relationships and laws. Both the lawfulness of the universe and the role of chance within this framework can be seen as God-given. Randomness brings forth potentialities written into the God-given universe.

Peacocke sees God as genuinely innovative and responsive to the evolving universe and to the evolution of life on Earth, constantly involved in exploring new possibilities, and working with the propensities built into the universe from the beginning. The Creator works creatively through the regularities of the laws of nature and through chance, exploring the potentialities inherent in the natural world and enabling the new to emerge.

If God's purposes for creation include the emergence of modern humans, then theology needs to think of God as possessing extraordinary patience. In the light of what we know from the sciences, God is to be thought of not only as empowering but also as waiting upon the long unfolding story that includes the 13.75-billion-year history of the expansion of the observable universe, the formation of galaxies and stars, the forging of elements like carbon in stars, the formation of our solar system around the Sun, the emergence of the first life on Earth 3.7 billion years ago, the evolution by means of natural selection of life in abundance, and of modern humans in the last 200,000 years.

John Haught sees the nature of the generous and humble Creator as entirely consistent with the fact that, to the eyes of science, the natural world can appear at the empirical level as making itself, through processes of emergence and evolution. Inspired in part by Alfred North Whitehead, Haught writes:

> Since it is in the nature of love, even at the human level, to refrain from coercive manipulation of others, we should not expect the world that a generous God calls into being to be instantaneously ordered to perfection. Instead, in the presence of the self-restraint befitting an absolutely self-giving love, the

---

[13] D. J. Bartholomew, *God of Chance* (London: SCM, 1984), 97.

world would unfold by responding to the divine allurement at its own pace and in its own particular way. The universe then would be spontaneously self-creative and self-ordering. And its responsiveness to the possibilities for new being offered to it by God would require time, perhaps immense amounts of it. The notion of an enticing and attracting divine humility, therefore, gives us a reasonable metaphysical explanation of the evolutionary process as this manifests itself to contemporary scientific inquiry.[14]

The evolutionary picture of the emergence of life is consonant with a Creator of self-giving love, the kind of love revealed in the cross. Love operates in a humble way in the evolutionary emergence of a universe of creatures. Love allows space and time for the beloved to emerge in its own proper way.[15]

In terms of the model I have been suggesting, it is the Word of God who is the Attractor, humbly drawing the whole creation into existence and identity. It is the Spirit of God who is the quiet, hidden Energy of Love who humbly enables and empowers this evolutionary emergence. It seems that the God of creation is a God of infinite patience who delights in processes that involve emergence though increasing complexity over long periods of time. Because of God's respect and love for finite creatures and processes, God accepts the limitations associated with creating through finite processes. The Christian vision of God's creative action in an evolutionary universe, then, is of Love that lives humbly with the process, accompanying creation in its emergence, delighting in its beauty and its diversity, suffering with it out of the divine passion of love, and promising its healing, liberation, and participation in deification in Christ.

---

[14] Haught, *God after Darwin*, 57.
[15] Ibid., 145.

Chapter 7

Neuroscience and the Creation
of the Human Soul

Thanks to fossil discoveries and modern genetics, we now possess a reasonably reliable account of the transition from the common ancestor we share with other primates to the various hominin species that began about six million years ago. We can trace the transition from various archaic species of *homo* to modern humans with their highly developed brains that took place about two hundred thousand years ago.[1] Recent work in neuroscience is giving us new insights into the way our highly evolved brain functions and its role in cognition and emotional life. How do these new evolutionary understandings of the brain relate to the Christian teaching concerning the human soul?

As a specific case of Christian teaching on the soul, I will consider the formal position of the Catholic Church articulated by two popes. I will trace the way that this church has moved toward an openness to, and acceptance of, the theory of evolution. But I will note that in this shift the popes express a reservation—the soul is immediately created by God. This teaching becomes an apparent sticking point for some Christians in relationship to the theory of evolution. It can seem to stand as an alternative understanding to that of science. In this chapter I will attempt to explore how the scientific view and the central understanding of the human person defended by the church might be seen as compatible with one another. I will begin by sketching something of what contemporary neuroscience says about the relationship between the brain and the mind. Then I will

[1] See Jerry A. Coyne, *Why Evolution Is True* (New York: Penguin, 2009), 190–220.

outline the standard Catholic teaching on the creation of the soul. In the third section I will offer a theological interpretation of the creation of the soul that aims to be, at least in broad terms, in creative dialogue with the findings of neuroscience.[2]

Neuroscience on the Brain and the Mind

As one admittedly limited way of attempting to enter into theological dialogue with contemporary neuroscience, I will focus on the work of Michael Gazzaniga. One reason for choosing Gazzaniga is that he is a leading figure in neuroscience who provides a helpful and widely read account of developments in the field.[3] Another reason is that he has a view of the brain-mind relationship with which theology can dialogue. He sees the mind as dependent on the physical brain but also as emergent upon the brain and constituting a new level of organization and control. He does not reduce the mind to the brain, nor see it as fully determined from below.

Early developments in neuroscience were often made by studying patients with damage to a particular part of the brain and observing its impact on particular capacities of the patient, such as those for sight, speech, the recognition of faces, and the emotional cues of others. The use of scanning techniques enabled researchers to show the relationship between activities in particular areas of the brain and specific cognitive functions. One result of this research has been the discovery that the brain is not the unified structure we tend to assume it is. It functions through a great many independent modules operating in parallel through neural networks, such as the visual network. These networks involve communications between populations of neurons distributed across different regions of the brain.

[2] I am grateful to Julie Clague for discussions that helped with the formulation of this section.

[3] Michael S. Gazzaniga is director of the SAGE Center for the Study of the Mind at the University of California, Santa Barbara, president of the Cognitive Neuroscience Institute, and director of the MacArthur Foundation's Law and Neuroscience Project. He delivered the Gifford Lectures at the University of Edinburgh in 2009.

In a popular book, Michael Gazzaniga describes his early work with "split-brain" patients.[4] In these patients, the connections between the two hemispheres of their brains had been cut in order to relieve severe epilepsy. The surgery reduced the patients' seizures, and they reported being otherwise largely unaffected. Working with these split-brain patients enabled Gazzaniga and others to explore the specific functions of the two sides of the brain. It became clear that the left hemisphere of the brain is where language and intelligence is centered. For this reason, split-brain patients could operate from the left hemisphere without loss of I.Q. But if an object was presented visually only to the right hemisphere, the patient had no words for it, and so reported seeing nothing. Amazingly, the patient was, nevertheless, able to draw the object. The right hemisphere is associated with detail, with the visual, the spatial, and the artistic.

The left brain–right brain distinction was a fundamental insight, but far more diversity was to be discovered. The brain contains many specialized modules performing discrete tasks operating in parallel and at the same time. These modules communicate across widely distributed networks involving both hemispheres. In spite of the left brain–right brain difference, and in spite of all the competing voices, we human beings still function with a sense of mental unity. What accounts for this? Gazzaniga's experiments indicate that there is a module of the left hemisphere that he calls the interpreter. The interpreter of the left brain takes whatever information comes to it and makes it into a coherent story. Even when information is lacking—because, for example, a split-brain prevents information coming from the right brain—the left brain interpreter still makes up a story. Our conscious awareness arises out of the left hemisphere's unrelenting quest to explain what has come to consciousness. The interpretation comes after the event. There is a time-lapse between things "popping" up into consciousness and the interpreter of the left brain making sense of them.[5]

Gazzaniga sees conscious awareness as like the tip of an iceberg. Below the water line are great numbers of unconscious activities going

[4] Michael S. Gazzaniga, *Who's in Charge? Free Will and the Science of the Brain* (New York: HarperCollins, 2011).

[5] Gazzaniga, *Who's in Charge?* 103.

on in our brains, including those that keep the heart pumping, bodily temperature regulated, and visual processing enabled. Below the water level too are unconscious positive and negative dispositions toward the world around us. Because of our evolutionary past, we are hardwired in preconscious ways, with tendencies, for example, to react to the rustle in the grass that might signal a snake, to build social coalitions, and to make negative moral judgments against cheaters.

Conscious mental states, Gazzaniga believes, "arise from our underlying neuronal, cell-to-cell interactions." But, he insists, these mental states "constrain the very brain activity that gave rise to them."[6] In Gazzaniga's view, brains and minds interact with one another. Beliefs, thoughts, and desires that arise from brain activity also act downward upon the brain, influencing decisions and actions. Gazzaniga proposes that, as we have learned to see the macroscopic world as emergent upon the quantum level of reality, so mental states may be seen as emergent upon the physical reality of the brain. He sees conscious thought, then, as an "emergent property." Consciousness is not reducible to the physical brain. For him, "mind is a somewhat independent property of the brain while simultaneously being wholly dependent on it."[7]

What of the human sense of freedom and responsibility? Gazzaniga finds the foundation for personal responsibility in the deeply social nature of the human brain and mind. We are wired from birth for social interactions. We learn by mirroring others. Much of what is unique to human cognition is constituted by social cooperation. In a way that has no parallel in other species, we have inherited the "theory of mind," the ability to read the desires, feelings, and intentions of others. We inherit moral intuitions that arise spontaneously in us far more quickly than our rational considerations of ethical matters. Gazzaniga suggests that these moral intuitions have been selected for in the coevolution of the brain with the human social context.[8]

---

[6] Ibid., 107.

[7] Ibid., 130.

[8] Ibid., 143–78. This is explained through the "Baldwin effect," the idea that while acquired characteristics cannot be inherited, the tendency to acquire certain traits can be (p. 153).

Gazzaniga recognizes that there is a deterministic view that surrounds much science, including neuroscience, promoting the bleak idea that we are no more than machines serving as vehicles for the physically determined forces of the universe. His alternative view is that mind is emergent on the brain, and that personal and social experiences impact upon our emergent mental systems, and are powerful forces that modulate our minds: "They not only constrain our brains but also reveal that it is the interaction of the two layers of brain and mind that provides our conscious reality, our moment in real time."[9] It is this interaction between the layers of mind and brain, and between persons in a social environment, that Gazzaniga sees as the place of mental life and of human responsibility and freedom.

Gazzaniga's position on mind as nonreductively emergent is not an isolated one. His former teacher, Roger Sperry, a founding figure in modern neuroscience and Nobel laureate, has said: "Consciousness is conceived to be a dynamic emergent property of brain activity, neither identical with nor reducible to, the neural events of which it is mainly composed."[10] Neuropsychologist Malcolm Jeeves has spoken of mental activity as dependent upon the physically determinant operations of the brain, but as far more than a mere epiphenomenon of this physical activity. An epiphenomenon is a secondary phenomenon that is caused by and accompanies a physical reality but has no causal influence itself. Jeeves sees mental activity as having its own causal activity. He understands the mind as *"embodied in* brain activity rather than as being *identical with* brain activity."[11] Along with philosopher Nancey Murphy and others, he advocates what he calls a "nonreductive physicalist" view of the relationship between the brain and the mind.[12]

---

[9] Ibid., 218.

[10] Roger W. Sperry, "Forebrain Commissurotomy and Conscious Awareness," in *Brain Circuits and Functions of the Mind: Essays in Honor of Roger W. Sperry,* ed. C. Trevarthen (Cambridge, UK: Cambridge University Press, 1990), 382.

[11] Malcolm Jeeves, "Brain, Mind and Behaviour," in *Whatever Happened to the Soul? Scientific and Theological Portraits of Human Nature,* ed. Warren S. Brown, Nancey Murphy, and H. Newton Maloney (Minneapolis, MN: Fortress Press, 1998), 89.

[12] See Nancey Murphy, "Human Nature: Historical, Scientific, and Religious Issues," and "Nonreductive Physicalism," in *Whatever Happened to the Soul?* 1–29,

Gazzaniga's neuroscience leaves some fundamental questions unanswered about the nature of mind, and I think that theologians may seek a stronger view of human freedom than he offers. But his empirical research provides brilliant new insights into the functioning of the brain, and its relationship to mind in an interactive social world.[13] His is a searching and open view of the mind as dependent upon the physical brain, but as an emergent reality that represents a new level of evolutionary complexity, the level of human consciousness and freedom in a social environment. Such a vision raises important questions for theology, particularly about the Christian notion of the soul and the Catholic teaching that God immediately creates the human soul.

Catholic Teaching on the Immediate Creation of the Soul

Granting the importance of the concept of soul in the Christian tradition, it is not surprising to see the Catholic Church defending the uniqueness of the human soul in twentieth-century discussions about evolution. It did this, in part, by teaching that the human soul is created immediately by God. In an important encyclical in 1950, Pope Pius XII supports prudent and moderate discussion of the scientific hypothesis of evolution "in as far as it inquires into the origin of the human body as coming from preexistent and living matter." Along with developments in biblical studies that better enabled a contextual and theological reading of Genesis rather than a literalist one, this teaching of Pius XII is an important step in this church's engagement with Darwinism. Until the 1950s, Catholic evolutionary thinkers like Teilhard de Chardin had faced active discouragement

---

127–48. See also Murphy's *Bodies and Souls, or Spirited Bodies?* (Cambridge, UK: Cambridge University Press, 2006).

[13] For a recent theoretical approach to the inclusion of meaning, purpose, and consciousness with evolutionary science, see Terence W. Deacon, *Incomplete Nature: How Mind Emerged from Matter* (New York: W. W. Norton & Company, 2011). Ideas similar to some of those found in this work were published earlier by Alicia Juarrero in *Dynamics of Action* (Cambridge, MA: MIT Press, 1999) and by Evan Thompson in *Mind in Life* (Cambridge, MA: Belknap Press, 2007).

from church leaders. The shift in the church's approach to evolution encouraged Catholic theologians to engage in a critically important dialogue with evolutionary thought in the second half of the twentieth century. But Pius XII went on to express a reservation about how this dialogue would affect Catholic understanding of the soul, saying that "the Catholic faith obliges us to hold that souls are immediately created by God."[14]

During his pontificate, Pope John Paul II gave his full support to the dialogue between theology and the sciences, and specifically to the theological discussion of biological evolution. On October 22, 1996, in a speech to the Pontifical Academy of Sciences, he spoke of developments in evolutionary science that had occurred since Pius XII's encyclical:

> Today, almost half a century after the publication of the encyclical, new knowledge has led us to realize that the theory of evolution is no longer a mere hypothesis. It is indeed remarkable that this theory has been progressively accepted by researchers, following a series of discoveries in various fields of knowledge. The convergence, neither sought nor fabricated, of the results of work that was conducted independently is in itself a significant argument in favor of this theory.[15]

This statement clearly signaled a positive approach to evolution. It recognized a scientific consensus that the church was obliged to take seriously. John Paul II's positive comments were widely dis-

---

[14] *Humani Generis*, 36. http://www.vatican.va/holy_father/pius_xii/encyclicals /documents/hf_p-xii_enc_12081950_humani-generis_en.html.

[15] John Paul II, "Message to the Pontifical Academy of Sciences (22 October 1996)," par. 4, in *Evolutionary and Molecular Biology: Scientific Perspectives on Divine Action*, ed. Robert John Russell, William R. Stoeger, and Francesco J. Ayala (Vatican City State / Berkeley, CA: Vatican Observatory Publications/Center for Theology and the Natural Sciences, 1998), 2. The important first sentence has been retranslated from the French by George Coyne (See n. 3, p. 5, and n. 15, p. 14, of *Evolutionary and Molecular Biology*). The French version of this sentence in *L'Osservatore Romano* (23 October 1996) reads: *"Aujourdhui, près dun demi-siècle après la parution de l'encyclique, de nouvelles connaissances conduisent à reconnaitre dans la théorie de l'évolution plus qu'une hypothèse."*

cussed not only in the media but also in scientific circles.[16] His view was welcomed and commented on by key evolutionary thinkers, including Stephen Jay Gould.[17] It gave renewed encouragement to theologians to take this evolutionary worldview seriously in their theological work.

John Paul II went on to quote the words of Pius XII about the immediate creation of the soul. He recast the discussion by emphasizing the different methodologies and ways of knowing in science and theology. But he insisted that the human person is irreducibly spiritual, and he found a purely materialist interpretation of the human person seriously deficient:

> Consequently, theories of evolution which, in accordance with the philosophies inspiring them, consider the spirit as emerging from the forces of living matter, or as a mere epiphenomenon of this matter, are incompatible with the truth about [the human]. Nor are they able to ground the dignity of the person.[18]

What is being rejected here is the reductionist interpretation that would see the spirit as a mere by-product of matter and biology, rather than as something that is essential to the constitution of the human person. Such a view would leave no room for God's creative act in the emergence of the human person, or for the idea of the human person's dignity grounded in the image of God. I do not interpret what is said here as in any way necessarily opposed to Gazzaniga's neuroscientific view of the mind as emergent upon the brain. On the contrary, I will propose that these views can be brought into a genuine dialogue. In fact, after pointing to the "ontological leap" that theology finds in the human person, John Paul II goes on to advocate a dialogue with the natural sciences:

[16] *The Quarterly Review of Biology* devoted an issue—number 72 (1997)—to the papal statement and to commentary on it by biologists, including Edmund Pellegrino, Michael Ruse, and Richard Dawkins.

[17] Stephen Jay Gould, "Nonoverlapping Magisteria," *Natural History* 106 (March 1997): 16–22; *Rocks of Ages: Science and Religion in the Fullness of Life* (New York: Ballantine Books, 2002).

[18] John Paul II, "Message to the Pontifical Academy of Sciences," par. 5, p. 6.

With [the human], then, we find ourselves in the presence of an ontological difference, an ontological leap, one could say. However, does not the posing of such ontological discontinuity run counter to that physical continuity which seems to be the main thread of research into evolution in the field of physics and chemistry? Consideration of the method used in the various branches of knowledge makes it possible to reconcile two points of view which would seem irreconcilable. The sciences of observation describe and measure the multiple manifestations of life with increasing precision and correlate them with the time line. The moment of transition to the spiritual cannot be the object of this kind of observation, which nevertheless can discover at the experimental level a series of very valuable signs indicating what is specific to the human being.[19]

The pope sees it as the role of philosophy to consider the human person's self-awareness, knowledge of God, moral conscience, and aesthetic experience. He sees it as the role of theology to show the ultimate meaning of all this in the light of the Creator. What John Paul II is suggesting is a dialogical stance, with science operating according to its empirical commitments, and philosophy and theology operating according to their own proper methodologies. It is my hope that this chapter will be a contribution to this dialogue. In the earlier part of this chapter, I have outlined some of the insights into the brain and the emergence of mental states offered by neuroscientist Michael Gazzaniga. In this section, I have represented the view of the immediate creation of the soul found in Catholic teaching. In the rest of this chapter, I will attempt to suggest how a theology of soul might be understood, at least in broad terms, in creative relationship to the insights of neuroscience.

A Theological Interpretation of Soul

The Christian tradition has been long committed to the idea that human beings are not only bodily but also spiritual. They are people of soul, people who have a spiritual soul. The idea of the

[19] Ibid., par. 6, p. 6.

soul includes the capacities that neuroscience speaks of as mind, or as mental states. Soul is the location of intellectual life, freedom, morality, and aesthetic experience. It is that center of our embodied selves from which we engage in interpersonal life and learn to love others. It is our inner selves, and it is also our outgoing personal selves engaged in a social, cultural world, and in relationship to, and dependent upon, the wider natural world of which we are a part. It is the self that responds to the presence of the Spirit of God. It is the inner place of mindfulness, prayer, and worship.

To say that soul is essential to the human person is not necessarily to advocate any one particular philosophical notion of the soul, whether Platonic, Aristotelian, or any other. It is certainly not necessarily an espousal of a dualistic notion of body and soul. For Christians, the sense of an ensouled self is grounded in the biblical idea of the whole person made in the image of God (Gen 1:27), known, called, and carried by God from her mother's womb to old age (Isa 46:3-4). Soul can be seen as the equivalent of what is called "heart" in the Bible. Heart is the place of the indwelling Holy Spirit: "God's love has been poured into our hearts through the Holy Spirit that has been given to us" (Rom 5:5). By this gift of the Spirit, Christians believe, we are made God's beloved daughters and sons, able to cry out "Abba! Father!" (Rom 8:15-17). As spiritual and bodily persons we are already being transfigured in Christ, the true Image of God—"being transformed into the same image from one degree of glory to another" (2 Cor 3:18). The God who said, "Let light shine out of darkness," has now "shone in our hearts to give the light of the knowledge of the glory of God in the face of Jesus Christ" (2 Cor 4:6). This promise involves our whole embodied selves, I believe, but it clearly involves what Scripture calls the heart and what theology calls the soul or the spirit.

What are we to make of the church teaching that our souls are immediately created by God in light of what neuroscience tells us about the evolution of the human? Are these ideas opposed, or can they be reconciled? I will suggest a theological interpretation of immediate creation that is open to a creative engagement with a neuroscientific account in six steps. These will build on the trinitarian theology of creation already discussed and make use of an insight from Karl Rahner.

1. *The unity of the one person.* The theological anthropology suggested here is that of the human as an irreducible unity of the whole person, body and soul. The person is understood as interrelational, in relationship with other persons, with the rest of the natural world and with the triune God. It is all too easy to interpret "the immediate creation of the soul" dualistically, as if body and soul are two separate, already-existing entities, as if God places a soul in an already-existing body. A concept of the unity of the whole person suggests that we are always embodied persons, as we are also always ensouled persons. Body and soul are both essential to the human person. Taking this view also suggests the theological position that eternal life is to be understood not as the ongoing existence of a separated soul, but as the participation, in a way beyond our comprehension, of the whole person, body and soul, in resurrection life. The triune God is Creator of the whole person, and the whole person, body and soul, is destined for life in the triune God.

2. *The unity of the one trinitarian act of creation.* God's creative action is one act of self-giving love, from the Father, through the Word, and in the Spirit. It is wonderfully differentiated in its outcomes in creatures, but it is the one act of the one triune God, rather than millions of separate acts. The unity of the one divine act of creation of a universe of creatures is grounded in the unity of the one God. This is why it is not helpful to interpret the "immediate creation of the soul" as if God acts in completely discrete, separate, and miraculous initiatives for each of the billions of human beings who have lived, are living, and will live. The immediate creation of the soul of each person is better understood as a unique and particular outcome of the one divine creative act, one act of the Father through the Word and in the Spirit, which embraces both original creation and the continuous creation of all things. This one act is differentiated and multiple in its outcomes in a world of finite creatures.

3. *Creation is directed toward deification.* There is an important theological tradition embraced here which argues that the liberating and transfiguring incarnation of the eternal Word is not

a divine plan B. It does not come about only because of sin. On
the contrary, God created a world of creatures in order to give
God's self to creation through the incarnation of the Word and
the gift of the Spirit. God always intended the incarnation, and
through it to bring creation to its liberation and deification in
the life of the Trinity. Creation is an act of divine self-bestowal
in love that is directed toward the incarnation of the Word
and the Pentecostal outpouring of the Holy Spirit, and to the
deification and final fulfillment of humans and with them of
the whole creation.

4. *God creates through enabling creatures to participate in the process
   through their own active self-transcendence.* The relationship of
   creation, by which God is present to all creatures through the
   Word and in the Spirit, not only enables creatures to exist and
   to interact within the community of creation but also enables
   created entities themselves to have the capacity for emergence—
   to become something new. The Holy Spirit is the indwelling
   Life-Giver enabling evolutionary emergence. The Word of
   God is the Attractor in this process, drawing creatures to their
   individuation within a community of creation and to their final
   fulfillment. The effect of the Spirit in creatures is that they
   themselves possess a capacity of emergence by means of natural
   processes and laws of nature. This applies in the evolution
   of modern humans from their hominin ancestors. Central to
   this process is the evolution of the human brain, especially its
   capacity for language and intellectual and interpersonal life.
   In and with this evolution of the brain in the human social
   world, the mind/soul emerges as profoundly connected to and
   dependent upon the brain but also as a new level of created,
   God-given reality in our world. The Spirit empowers this emergence from within, the Attractor draws human beings to their
   personal and communal identity in the triune God.

5. *God immediately creates each soul.* Athanasius teaches us that
   God creates every entity, every sparrow, every worm, and every
   whale through the Word and in the Spirit. He insists that this
   act of continuous creation is immediate to all things. It is God's

immediate presence that enables each creature to partake of the Word in the Spirit. There can be nothing more immediate than this trinitarian presence. So in a fundamental sense, the immediacy of God's creative presence applies to all creatures, and not just to the human soul. God immediately creates everything, including the human soul. But clearly something more is being suggested when it is said that God immediately creates the soul. I think that this "more" involves not just the immediacy of God, but this immediacy directed toward the particular outcome of this specific ensouled person.

6. *God's creative act has this person's soul as a specific, directly willed outcome.* God's creative act is directed to this particular person, with her own personal spiritual identity, uniquely created according to the Image that is the Word. As Karl Rahner has said, the idea of immediate or direct creation does not exclude God acting through evolutionary processes, as long as we see that something new and independent is brought into existence by God in such a way that God's creative act "directly terminates" in the creature.[20] God's one creative act has this person and her personal soul as the divinely willed outcome. She is known from her mother's womb, called by name to be filled with the Spirit, deified in Christ and carried through life and death as God's beloved daughter. Her ensouled personal self is God-given, the gift of the Spirit, uniting her to the Word, and through the Word to the Father. This one creative and grace-full act of God the Trinity, with its unique and specific outcome in each person, can be understood as an appropriate theological interpretation of what is meant by the church teaching of the immediate creation of the human soul.

My proposal in this chapter is not that neuroscience and theology are saying the same thing, or that one necessarily implies the other.

---

[20] Karl Rahner, "Evolution," in *Encyclopedia of Theology: A Concise Sacramentum Mundi*, ed. Karl Rahner (London: Burns and Oates, 1975), 488. See also Rahner's *Hominisation: The Evolutionary Origin of Man as a Theological Problem* (New York: Herder and Herder, 1968).

It is simply meant to suggest: 1. that they are not mutually opposed; 2. that there is room for ongoing dialogue between them; 3. that it is entirely possible to hold both a neuroscientific view of the brain and a theological view of the human soul, including the idea of immediate creation. What is required is an interpretation of neuroscience of a nonreductive and open kind and a theological interpretation of the creation of the soul that is theologically plausible within an evolutionary perspective.

Evolution, Cooperation, and the Theology of Original Sin

Does the Christian doctrine of original sin have any meaning in an evolutionary view of the world? The proposal I am advancing here is that the concept of original sin remains indispensable and that it can have new meaning in an evolutionary context. The chapter begins with a brief discussion of original grace which will set the context for exploring original sin. I then take up some of the insights offered by recent scientific work on the evolution of cooperation. This will lead to a reflection on grace and sin within this new context, leading to some conclusions on original sin.

Original Grace

One of the great advances in Christian theology during the twentieth century is the new clarity with which major Christian churches teach that God's saving grace is not limited to Christians. Grace is universal in its reach. The salvation given in Jesus Christ, the Word made flesh, is not restricted to Christians. It is offered to all. Grace is at work in all human lives, a grace each human can freely embrace or reject.

The Second Vatican Council was quite explicit. In *Lumen Gentium* (the Dogmatic Constitution on the Church) we find: "Those who, through no fault of their own, do not know the Gospel of Christ or his church, but who nevertheless seek God with a sincere heart, and, moved by grace, try in their actions to do his will as they know it through the dictates of their conscience—these too may attain eternal salvation" (LG 16).[1] In *Gaudium et Spes* (Pastoral Constitution on the Church in the Modern World), the Council speaks of the Christian belief that salvation comes through Christ's paschal mystery—his death and resurrection and the outpouring of the Holy Spirit. The text goes on to insist that saving grace is made available, in a way known only to God, to all people of good will:

> All this holds true not only for Christians but also for all people of good will in whose hearts grace is active invisibly. For since Christ died for everyone, and since all are in fact called to one and the same destiny, which is divine, we must hold that the Holy Spirit offers to all the possibility of being made partners, in a way known to God, in the paschal mystery. (GS 22)

The Holy Spirit is present to every human being, offering salvation to each. Earlier in this book I explored salvation as a partaking of Christ, through the Holy Spirit—a deifying participation in the life of the Trinity. But Christian faith can recognize that a person from another religious tradition, or with no religious affiliation, may experience the presence of the Spirit, and in their own way say yes to this gift of God, without ever naming it in a Christian way. Sometimes the spirit is present quietly in the experience of deep peace. Sometimes the spirit is present in a challenging, disruptive way. The Spirit is there in each life, for each person, sometimes like a mighty whirlwind (Job 38:1), sometimes like the "sound of sheer silence" (1 Kgs 19:12). In the Holy Spirit, Love comes seeking each of us. Our God-given freedom means that each person is free to accept this saving, transforming Love, or to reject it.

To understand the work of the Spirit, then, we need to go well back—beyond the last two thousand years, beyond Moses, beyond Abraham and Sarah, beyond the Neolithic revolution, beyond the movement of modern humans out of Africa. As Pope John Paul II said in his encyclical on the Holy Spirit (*Dominum et Vivificantem*), "*We need to go further back*, to embrace the whole of the action of the Holy Spirit even before Christ—*from the beginning*, throughout

---

[1] *Vatican Council II: The Basic Sixteen Documents*, trans. Austin Flannery, OP (Northport, NY: Costello Publishing, 1996), 367.

the world, and especially in the economy of the Old Covenant."[2] How might we think about this universal presence of the Spirit in the light of human evolutionary history? Earlier in this book I proposed that the whole process of the emergence and expansion of the observable universe over the last 13.75 billion years has been the place of the nearness of God, with the Word of God as the Attractor of its emergence and the Spirit as the Power of Love. If this is so, then the universe has always been the place of God. It has always been grace-filled. But Christians see grace in a particular way as God's self-offering in love to human beings, to those in whom the universe comes to consciousness, freedom, and personal love.

When human beings evolved on Earth, they emerged into a grace-filled universe. They emerged into the presence of the Spirit, Love directed to them, freely offered to them. We know that the emergence of the human involves a process that began about 6 million years ago, when the line of hominins diverged from a common ancestor with other apes. It continued in the australopithecines, then in various species such as *Homo erectus*, and later in archaic human species like the Neanderthals and in modern humans who evolved within the last two hundred thousand years.[3] We do not know precisely when humans became capable of religious experience and of some form of particularly human response to Love's offer. But the theological claim is that when this occurred they emerged into a graced world. They were surrounded and upheld by the Spirit, present to them as self-offering divine love, constantly inviting them to embrace what was freely offered.

Original Sin

I have been suggesting that we can see the grace of the Spirit as present in our human origins. Grace is original. What of sin? How might we think about the doctrine of original sin within an evolutionary perspective? Original sin is usually taken to mean that there

---

[2] John Paul II, *Dominum et Vivicantem*, 53. Translated as *The Holy Spirit in the Life of the Church* (Boston: St Paul, 1986), 91.

[3] See Jerry A. Coyne, *Why Evolution Is True* (New York: Penguin, 2009), 190–220.

was a sin against God, a rejection of God's grace, in human origins. This rejection involves the whole human race in alienation and brokenness, but God's action in Christ brings forgiveness, healing, liberation, and deification. According to this tradition, solidarity in original sin is transmitted by propagation—it is something we inherit. The result of original sin is a tendency to sin, a pull toward evil, in each of us. The word sin is used analogously. Original sin is not freely chosen personal sin, but a tendency to sin that coexists in us with the gracious presence of God in self-offering love.

My impression is that some contemporary Christians are inclined to abandon this doctrine as meaningless or as an oppressive relic of the past. They take this position for a variety of reasons: they have adopted an evolutionary view of life on Earth; they see biological death as one of the costs of evolution, rather than as the result of human sin; they interpret Genesis not in a literal fashion, but contextually, figuratively, and theologically; they do not see Adam and Eve as historical figures; they do not see the story of their disobedience as the literal description of a historical event; they see the description of God's actions in Genesis 2–3 as anthropomorphic; they oppose the way Eve's part in the story has been used to justify misogyny; they are critical of the way some theologians have linked the transmission of original sin to a negative view of human sexuality; they reject the idea, held until recently by many in the Catholic Church, that because of original sin unbaptized children would not find eternal life in God.

Each of these critical responses is appropriate and justified. Any modern articulation of original sin needs to acknowledge them. But what is needed, I believe, is not an abandonment of original sin, but a renewed and deepened theology of original sin that takes each of these critical positions into account. Above all it needs to be a doctrine that embraces biological evolution.

Why should we make the effort to continue to take this doctrine seriously? One reason is that original sin has been understood as central to Christian faith and deeply connected to salvation in Christ. In the Catholic Church it is taught by councils of the church, particularly the Council of Trent. This suggests that Christians today might do well to examine rather than ignore this doctrine, asking what enduring meaning it might have in a new context.

Is there a wisdom at the heart of the doctrine of original sin that we need, a wisdom that our world needs? When the question is asked in this way, I think an answer starts to emerge. Original sin brings a strongly critical note to the theological view of the human person. It reminds us not only that human beings are graced but also that they have the capacity for terrible evil. The theology of original sin challenges all romantic views of humanity, all uncritical views of evolutionary progress, all idealistic visions of the future, all naïve revolutionary utopias. It is a doctrine of radical realism about humanity. Such a critical note is not opposed to vision and hope, I believe, but grounds them in truth. In this chapter I will not attempt a full theology of original sin. Rather, I wish to simply explore one of the ways I see original sin as operative and therefore deeply meaningful in a modern context: namely, the tendency to make others into scapegoats and enemies.

Human beings can bond together and create intimacy by making others into victims of exclusion or violence. Children gang up on a vulnerable child in a school playground. Adolescents bully one another in the locker room or in cyberspace. Adults build connections with one another by gossip that demeans, excludes, and damages others. Populist politicians seek to bond with voters by attacking "outsiders." War seems always to fill the pages of our newspapers and television screens with stories of the killing of noncombatants, of rapes, of torture, of the maiming of children, and of the terrible loss of life and limb suffered by combatants. We live still with the horrors of the Second World War, the firebombing of great cities and their defenseless civilian populations, Hiroshima and Nagasaki, and the extreme of human violence, the systematic murder of six million Jewish people in the Shoah. We human beings seem to have an endless capacity to make other people into a common enemy and to unleash violence against them. In many instances, we commit extreme atrocities unknown in the rest of the natural world. It has been estimated that the conflicts and wars of the twentieth century killed 231 million people.[4]

---

[4] Milton Leitenberg, "Deaths in Wars and Conflicts in the 20th Century," http://www.cissm.umd.edu/papers/files/deathswarsconflictsjune52006.pdf. Accessed October 9, 2013.

Anthropologist René Girard has analysed the human tendency to violence in terms of the "scapegoat mechanism."[5] He sees this as grounded in rivalry that flows from our imitation (*mimesis*) of others—we want what others want. Violence can erupt from a conflict of desires, but it is contained by rivals uniting against a common enemy: "Suddenly the opposition of everyone against everyone else is replaced by the opposition of all against one. Where previously there had been a chaotic ensemble of particular conflicts, there is now the simplicity of a single conflict: the entire community on one side, and on the other, the victim."[6]

Raymund Schwager and other theologians have taken up the idea of the scapegoat mechanism as a central theme in a renewed theology of original sin and salvation. In Schwager's view, Jesus dies the death of a scapegoat, and in his death and resurrection the Spirit is poured out, liberating us from scapegoating and constituting the beginning of a radically inclusive new creation. Before his death, Schwager had begun to explore scapegoating in relation to our evolutionary inheritance, seeing sin as "woven into the natural tendencies of human life" through the course of evolutionary history.[7] I will seek to take some of these ideas further, asking what recent work on the evolution of cooperation might suggest about grace and sin.

The Evolution of Cooperation

One of the important intellectual developments of recent decades has been in the biological understanding of the evolution of

[5] See René Girard, "Violence, Scapegoating and the Cross," in *The Evolution of Evil*, ed. Gaymond Bennett, Martinez J. Hewlett, Ted Peters, and Robert John Russell (Göttingen: Vandenhoeck & Ruprecht, 2008), 334–48; *Violence and the Sacred* (Baltimore: Johns Hopkins University Press, 1977); *Things Hidden Since the Foundation of the World* (Stanford: Stanford University Press, 1987); *The Scapegoat* (London: Athlone Press, 1994); *I See Satan Fall Like Lightning* (Maryknoll, NY: Orbis, 2001).

[6] Girard, *Things Hidden*, 24.

[7] Raymund Schwager, *Banished from Eden: Original Sin and Evolutionary Theory in the Drama of Salvation*, trans. James G. Williams (Herefordshire: Gracewing, 2006), 55.

cooperation. I will seek to represent this development, in a partial way, with some key insights from the experimental work of Michael Tomasello and the more theoretical ideas of Martin Nowak. I see this as a kind of snapshot of a much wider field of work in progress from which theology can learn. It is important to acknowledge that this is a developing area of research, with its own controversies, and the different views of Tomasello[8] and Nowak[9] are challenged by other specialists.

Psychologist Michael Tomasello is the codirector of the Max Planck Institute for Evolutionary Anthropology in Leipzig. In a recent book, he asks why it is that human beings cooperate with one another. He recognizes, of course, that humans can be selfish, highly competitive, and exploitative of others. But from an evolutionary perspective, these attributes can easily be explained by natural selection. What explains the fact that human beings also seem highly cooperative when compared to their nearest primate relatives?

There are times when humans help others at great cost to themselves, including the sacrifice of their lives. In more ordinary ways, human beings help one another by cooperating in a wide range of large-scale undertakings such as agriculture, industry, commerce, politics, education, and religion. A moment's thought reveals the multilevel cooperation that is involved in such ordinary events as buying fresh food in a store or driving a car across a city. Humans live

---

[8] See Michael Tomasello, with Carol Dweck, Joan Silk, Brian Skyrms, and Elizabeth Spelke, *Why We Cooperate* (Cambridge, MA: The MIT Press, 2009), where Tomasello's four co-authors offer alternative views on some issues: Joan Silk sees altruistic tendencies as forming the basis for cooperation (111–22); Carol Dweck stresses the importance of learning rather than inheritance in the development of cooperation in infants (125–34); Brian Skyrms suggests that complex cognitive developments may not be necessary for cooperation, pointing to social insects and bacteria (137–46); Elizabeth Spelke asks whether language is more fundamental than the capacity for cooperation (149–72).

[9] When Nowak published an article with Edward O. Wilson and Corina E. Tarnita, "The Evolution of Eusociality," in *Nature* 466 (26 August, 2010): 1057–62, arguing that "inclusive fitness," or kin selection, is not necessary to explain cooperation in social insects or human beings, it was followed by critical letters, one of them containing the signatures of 137 scientists.

by everyday social norms that mandate, for example, the way they wait in a line for a bus, pay for goods taken in a store, and respect the privacy of others. When they fail to obey these norms, they can feel guilt and even face the possibility of punishment from others.

Tomasello and his colleagues have sought to understand cooperation through a series of comparative experiments that involve young children, on the one hand, and chimpanzees and other primates, on the other. These experiments have shown the surprising degree to which very young children help, inform, and share with others. Infants from fourteen to eighteen months old are found regularly to help others, for example, by fetching an out-of-reach object, or by opening a cupboard door for an adult whose hands are full. They consistently inform others, for example, of the whereabouts of objects by pointing, something that chimpanzees never do as a way of offering information. They share things, such as food, often equitably, in a way that chimpanzees do not. Young children discern other people's intentions by following the direction of their gaze. They not only practice but also expect a level of social conformity in games. Studies have found that three-year-old children not only participate in the norms of games but also insist upon them and enforce them.[10]

Tomasello proposes that from a young age, children possess a "social rationality" and a "shared intentionality." They are prepared to participate with others in a common goal that creates an "us."[11] He recognizes, of course, that children can be selfish in many situations. But what accounts for their altruism? He sees altruism as emerging not only from learning but also from the social nature of young children. In his view, the tendency to cooperate is not only culturally learned but also innate. Children come to culture with already-existing tendencies to be helpful, informative, and generous. In fact, Tomasello points out, cultural engagement can diminish children's tendency to cooperate. They observe whether others take advantage of them or reciprocate, and this leads them to be more selective about their altruism.

[10] Tomasello, *Why We Cooperate*, 37.
[11] Ibid., 40–41.

Chimpanzees help one another in some situations, and Tomasello sees in this an evolutionary link to the human tendency to cooperate. But he finds that chimpanzees do not cooperate in shared goals. Even in the group activity of the hunt, he argues, chimpanzees do not work together in a collaborative way toward a mutually beneficial outcome—"each participant is maximizing its own chances of catching the prey, without any prior joint goal or plan of assignment of roles."[12] They still function in an "I" mode. It is worth noting in passing, however, that some specialists in the behavior of apes see them as highly cooperative. Frans De Waal offers many examples of emotional bonds, cooperation, joint action, and conflict resolution among chimpanzees and bonobos. De Waal sees our human moral behavior as having an evolutionary basis and as deeply rooted in our primate legacy.[13]

Tomasello finds that the great apes and young children differ in three sets of psychological processes: in coordination and communication, in tolerance and trust, and in norms of cooperation and conformity. Human beings live by cooperative, institutional agreements, and these begin in children's play. Their joint agreement, for example, to treat a stick as a horse during playtime can be understood as a precursor to the symbolic and institutional agreements whereby adults see this paper as money or this person as president.[14]

In Tomasello's view, altruism and culture are born from the human capacity for cooperation that is already evident in very young children. While he says that we do not know how and why cooperation arose in human evolution, one speculation is that, in hunting and gathering food, "humans were forced to become cooperators in a way that other primates were not."[15] In his view, humans have

---

[12] Ibid., 63.

[13] Frans De Waal, *The Bonobo and the Atheist: In Search of Humanism among the Primates* (New York: W. W. Norton, 2013).

[14] Tomasello, *Why We Cooperate*, 97.

[15] Ibid., 99. Samuel Bowles and Herbert Gintis propose that it is the distinctive human capacity for institutional behavior and for the cultural transmission of learned behaviors that enabled the creation of a social and biological niche favorable to cooperation: "Our ancestors used their capacities to learn from one another and to transmit information to create distinctive social environments. The

evolved a species-specific capacity for social relationships and for shared intentionality. They have a natural tendency to cooperate and work together in a common goal. Tomasello notes that human cooperation tends to be inside the group, while heinous crimes can be committed against those who are outside the group.[16]

Martin Nowak does a very different kind of research on cooperation. He is a theoretical biologist, professor of biology and mathematics at Harvard University and director of the Program for Evolutionary Mathematics. He works with mathematical approaches to evolution and makes use of game theory, particularly the Prisoner's Dilemma, which models the struggle in life between defection and cooperation. Natural selection suggests a strong tendency to defect, to put oneself first, rather than to cooperate. What Nowak's models show is that cooperation can win out over defection, but that there is always an unstable tension between selfishness and cooperation. He sees cooperation as built into evolution from the beginning, as "woven into the very fabric of the universe."[17] But how does cooperation emerge in a Darwinian world? What are the mechanisms for the emergence of cooperation?

Nowak proposes that five basic evolutionary mechanisms enable cooperation: direct reciprocity (you scratch my back and I will scratch yours); indirect reciprocity (doing the right thing earns me a better reputation); spatial mixing of populations (clusters of cooperators can prevail even in the midst of defectors); group selection (groups with more cooperators outcompete other groups); kin selection or "inclusive fitness" (cooperation can emerge more easily

---

resulting institutional and cultural niches reduced the costs borne by altruistic co-operators and raised the costs of free-riding. Among these socially constructed environments, three were particularly important: group-structured populations with frequent lethal intergroup competition, within-group levelling practices such as sharing food and information, and developmental institutions that internalized socially beneficial preferences." *A Cooperative Species: Human Reciprocity and its Evolution* (Princeton: Princeton University Press, 2011), 197.

[16] Tomasello, *Why We Cooperate*, 99–100.

[17] Martin A. Nowak, with Roger Highfield, *SuperCooperators: Altruism, Evolution, and Why We Need Each Other to Succeed* (New York: Free Press, 2011), 16.

among closely related individuals).[18] Nowak sees group selection, in a coevolution of genes and culture, as playing a particularly important role in human evolution, operating together with direct and indirect reciprocity. His is a multilevel selection theory.[19]

More broadly, Nowak sees cooperation as underpinning all the innovations of evolution. It is the "architect of creativity throughout evolution, from cells to multicellular creatures to anthills to villages to cities."[20] Without cooperation there would be no construction and no complexity in evolution. He makes the strong claim that cooperation must now be considered as the third basic principle of evolution alongside mutation and selection:

> Previously, there were only two basic principles of evolution—mutation and selection—where the former generates genetic diversity and the latter picks the individuals that are best suited to a particular environment. For us to understand the creative aspects of evolution, we must now accept that cooperation is the third principle. For selection you need mutation and, in the same way, for cooperation you need both selection and mutation. From cooperation can emerge the constructive side of evolution, from genes to organisms to language and complex social behaviors. Cooperation is the master architect of evolution.[21]

Nowak sees cooperation as functioning before life itself begins, in the operation of the RNA molecules that give rise to life. He discusses in detail cooperation at work in communities of cells, in ant colonies, and then in the evolution of human language and human societies. He sees language as central to the evolution of the human, and language and cooperation as coevolving.[22]

---

[18] Ibid., 21–111.

[19] Sarah Coakley has engaged closely with Nowak's work and developed a renewed theology of sacrifice and a renewed natural theology in her 2012 Gifford Lectures, *Sacrifice Regained: Evolution, Cooperation and God*. http://www.faith -theology.com/2012/05/sarah-coakley-2012-gifford-lectures.html.

[20] Nowak, *SuperCooperators*, xvii.

[21] Ibid., xviii.

[22] Ibid., 173.

Human beings can be seen as "SuperCooperators," who use all five mechanisms of cooperation.[23] Nowak's calculations suggest that, in face of nature's competitiveness, the successful strategies necessary for human coexistence must have the "charitable" attributes usually associated with the religions of the world. To coexist peacefully and cooperate well, human beings need to be "hopeful, generous, and forgiving."[24] Above all, he insists, we need these attributes today if we are to deal cooperatively with the climate emergency that we face as a human community. Nowak believes that climate change will force the human community on Earth to enter into a new stage of cooperation.[25]

Grace and Sin in Evolutionary Context

What does all this mean for a Christian anthropology? Key ideas that I want to pick up from this reflection on the work of Tomasello and Nowak are these: (1) While there is a tendency in human beings toward self-interest, there is also a tendency toward cooperation. (2) Both these tendencies are innate, part of our genetic as well as our cultural heritage. (3) Cooperation is not simply an option for human beings, but intrinsic to human evolution, to human nature, and to human culture. (4) The evolutionary tendency toward cooperation and altruism is directed toward insiders and against outsiders.

Nowak draws attention to the connection between what effective cooperation demands and the virtues taught by world religions— being hopeful, generous, and forgiving. It would be an easy step in a Christian theology to move from this insight to the idea that our innate tendency to cooperate is an unalloyed good; it would be a simple step too to see the evolutionary tendency toward self-interest as evil. Cooperation would then be aligned with grace, self-interest with sin. I think, however, that this would be a serious mistake.

On the one hand, while the tendency to cooperate, in theological terms, is part of God's good creation, it is not yet virtue or grace.

[23] Ibid., 275.
[24] Ibid., 272.
[25] Ibid., 278.

It needs to be transformed by grace to become virtue. In some instances, cooperation without grace can be dangerous. We can cooperate in evil, as well as in good. Clearly cooperation in torture or in economic exploitation of the poor is evil. Furthermore, the tendency to cooperate can end in mindless and irresponsible conformity, as it did in an extreme way in Nazi Germany. This, of course, is a serious danger for all kinds of groups, including religious ones.

On the other hand, self-interest is not necessarily to be associated with evil. At a fundamental level, it is the tendency we inherit, by way of natural selection, which directs us toward seeking our own survival and generativity. Theologically, this tendency too can be seen as part of God's good creation. It is a tendency that is open to grace and can become virtue. The tendency to self-interest can enable us to resist conformity. It can enable us to refuse to be doormats for others. More positively, it can enable us to claim our own personhood, with our own identity, dignity, moral conscience, and integrity before God.

What I am suggesting is that human beings can be open to the Spirit of God in both cooperation and in self-affirmation, and that both of these also have the potential to become the place of sin. For Christian theology, other-orientation and self-affirmation together can be seen as a fundamental way in which the human being is made in the image of God. The God whose image we reflect in our limited human way is Communion-in-Love. In the Communion of the Trinity, the divine persons possess the fullness of distinctive personhood precisely in the unity of love. In this kind of unity and "cooperation," personal distinction is not lost, nor swallowed up, but is revealed as the eternal, primordial reality of God. The Christian vision of the human person, then, is that of one who is interpersonal by nature but who precisely in relationships with others is constituted fully in her own personal being.

Grace is the wordless presence of the Spirit, the generous presence of Love, to each person from the beginning. Each is drawn by the Word to truly find herself in going out of herself in love, learning to become generous, loving, and forgiving parents, friends, companions and collaborators. In the journey of our lives, the tendencies of our nature, toward the other and the self, can be transformed by the grace of the Spirit and become virtue.

Original Sin: Insiders and Outsiders

What, then, of original sin? My proposal is that one important part of original sin can be seen as the inherited tendency to make others into outsiders and enemies, even as we form cooperative communities of insiders. This is not a complete theology of original sin, and that other suggestions have been made about evolution and original sin is apparent in the theological literature.[26] But I think there is a fundamental rejection of God, which is part of human origins, in the choice made to build alliances at the expense of making others into scapegoats and enemies. The result is what biologist Edward O. Wilson calls our human "tribalism." He sees tribalism, the human tendency to form groups and to defend them strongly against other groups, as among the "absolute universals" of human nature.[27]

The emergence of ethical behavior that extends love, compassion, or help to outsiders is not favored by our heritage. The tendency that we inherit is for cooperative and altruistic behavior toward insiders. But a genuine ethics has to reach beyond insiders. As evolutionary biologist Ernst Mayr has said, real ethical behavior requires a *transformation* of our evolutionary inheritance, "a redirecting of our inborn cultural tendencies toward a new target: outsiders."[28] A new cultural factor is required for such an ethical transformation toward

[26] Some stress selfish tendencies inherited from prehuman ancestors. See Daryl P. Domning, with Monica Hellwig, *Original Selfishness: Original Sin and Evil in the Light of Evolution* (Aldershot, UK: Ashgate, 2006). Others stress the dissonance between our inheritance and culture. See Gerd Theissen, *Biblical Faith: An Evolutionary Approach* (London: SCM, 1984); Philip Hefner, *The Human Factor: Evolution, Culture, and Religion* (Minneapolis, MN: Fortress, 1993). Others refer to the "fall upwards" of human culture. See Jerry D. Korsmeyer, *Evolution and Eden: Balancing Original Sin and Contemporary Science* (Mahwah, NJ: Paulist, 1998); Marjorie Hewitt Suchocki, *The Fall to Violence: Original Sin in Relational Theology* (New York: Continuum, 1994).

[27] Edward O. Wilson, *The Social Conquest of Earth* (New York: Liveright Publishing Corporation, 2012), 57. As a Christian theologian seeking to engage with evolutionary thought, I am disappointed to find that after his earlier appeal for collaboration with religions in dealing with the ecological crisis in *The Creation: An Appeal to Save Life on Earth* (New York: W. W. Norton, 2006), Wilson ends this recent book by rejecting any possibility of reconciliation between science and the teachings of organized religions (p. 294).

[28] Mayr, *What Evolution Is* (London: Weidenfeld & Nicolson, 2001), 259.

outsiders. Mayr suggests that we need something like the teaching of a great philosopher or the preaching of a great prophet to move us to include outsiders within the range of our ethical concerns and actions.

Christians see this cultural factor in God's self-revelation in Israel and its culmination in Jesus Christ. It finds expression in his priority for the poor, his healing ministry, his celebratory meals with outsiders, and, above all, in his radical teaching of love for the enemy: "Love your enemies, do good to those who hate you, bless those who curse you, pray for those who abuse you" (Lk 6:27; Matt 5:44). The Christian claim is that what happens in the Word made flesh is not only a matter of new teaching and new praxis but also a divine act of new creation. In the life, death, and resurrection of Jesus, and the outpouring of the Holy Spirit, there is a divine act of deification at work in our world, which means that, in principle, human nature itself is already being transformed.

The Good News of Christianity is that in Jesus Christ, in the power of the Spirit, we find a transformation of our evolutionary inheritance that brings liberation from scapegoating and enemy-making. The divine defenseless love poured out upon Earth from the cross changes everything. In the power of the Spirit, the Christian community witnesses to this new reality that has broken in on history and has begun to transform it from within. In some way the Christian community embodies this new creation. But there is a real problem here, because we Christians find the same old evolutionary dynamics at work in the life of the Christian church, and in our personal and professional lives: us and them, insiders and outsiders. The Christian community carries the message of the Gospel of Jesus and embodies the transforming, liberating Spirit, but does this as a human community still subject to tendencies to exclude others and make them into scapegoats and enemies.

Evolutionary biologist David Sloan Wilson, who studies religions from a biological perspective, offers some important observations on these issues. In his work *Darwin's Cathedral*, he argues that organized religions are not to be seen as cultural parasites, maladaptive relics of the past or unimportant byproducts of evolution.[29]

---

[29] David Sloan Wilson, *Darwin's Cathedral: Evolution, Religion, and the Nature of Society* (Chicago: University of Chicago Press, 2002). Wilson's views have their

On the contrary, they can be highly functional organisms that attain the evolutionary goals of survival and reproduction. Because self-interest acts against altruism, accounting for what supports altruism has been a key issue for evolutionary theory, and Wilson argues that religions provide an important part of the explanation. He proposes that religions offer mechanisms of social control, at low cost to the individual, that support altruistic behavior within a group. They can promote self-sacrificing behavior, not only through effective social control mechanisms, but also through their belief systems that motivate group-benefitting behavior. In evolutionary terms, this is highly adaptive. Such behavior enables a community to survive and reproduce.

Wilson examines Calvinism, the Water Temple faith of Bali, Judaism, and early Christianity in particular. He shows how, in each case, religions enable communities to overcome self-serving egoism and to operate in cohesive and altruistic ways which achieve practical, secular outcomes. He finds, for example, that Calvin's church "included a code of behaviors adapted to the local environment, a belief system that powerfully motivated the code inside the mind of the believer, and a social organization that coordinated and enforced the code for leaders and followers alike."[30] From a biological point of view, religions, then, can have a positive role in enabling human survival and flourishing. A biological explanation of religion as adaptive, in my view, can happily coexist with a theological explanation of religion as, for example, divine self-revelation—as long as one accepts the idea of different levels of explanation for the same reality.

It must be noted, however, that the analysis Wilson offers has its dark side. The more cohesive and effective religious groups tend to be demanding of their participants. They require clear boundaries. Historically, this has resulted in serious conflicts between religions, and in religions engaging in exclusion, repression, persecution, and violence. And while some religions, and some religious leaders and

---

opponents. For a wider range of views on the biological origins of religion, see *The Believing Primate: Scientific, Philosophical, and Theological Reflections on the Origin of Religion*, ed. Jeffrey Schloss and Michal J. Murray (Oxford, UK: Oxford University Press, 2009).

[30] Ibid., 111.

prophets, have strongly opposed war, religions often function to support warring states. Religions, including Christianity, are not certainly immune from enemy-making. As Wilson says, "Religions are well known for their in-group morality and their out-group hostility."[31]

Christianity has at its core the good news of a divine love that shatters all boundaries, that demands of us nothing less than love for the enemy. But this good news is carried in a human community that functions in the way human groups do—with the tendency to set boundaries, to exclude the other. The church shares with other groups the evolutionary tendencies toward insider-outsider attitudes. Paradoxically, the more committed it is to the message of Jesus who ate with outsiders and sinners, the more exclusive it can tend to be. Even within a local church we form insider/outsider coalitions, between liberal and conservative, committed and non-committed, sophisticated and popular.

All of this is, I think, part of what the tradition has called original sin. In Christ, we can hope for and receive the gift of a new, more inclusive and generous love. Even now we experience this in partial ways in ourselves and in our communities, in moments of grace, in breakthroughs to the other when we can recognize the Holy Spirit at work. But to be open to such grace requires of us that we also face the truth of our wonderfully gifted but still dangerous human nature in true humility.

[31] Ibid., 10.

Part Three

Ecological Conversion

Human actions have put our planet, with its topsoil, forests, rivers, seas, and atmosphere, under increasing stress. We face an accelerating loss of biodiversity and extremely dangerous climate change. Much damage has already been done, and much is lost. But there is also a growing human response, a movement of diverse people connected in a deepening commitment to the good of the community of life on Earth. These are people who are undergoing a change of mind and heart to a deeper respect for other species, a commitment to protect them and enable them to flourish, along with the forests, the rivers, the seas and land.

In the last section of this book, I will explore this change of mind and heart in terms of the theme of ecological conversion, which I find in Pope John Paul II and in his common teaching with Bartholomew I, the Ecumenical Patriarch of the Orthodox Church. It is important to acknowledge the strong positions taken in teaching and in prophetic action by Protestant churches and the World Council of Churches, as well as by Orthodox and Anglican leaders. But I will limit myself here to the attempt to develop the theme of ecological conversion in the light of the theology articulated in this book.

Ecological Conversion in Church Teaching

The foundational text in Catholic ecological teaching was John Paul II's statement for the World Day of Peace, of January 1, 1990, entitled *Peace with God the Creator: Peace with All Creation*. In it he describes the growing ecological awareness as something that must be recognized and encouraged so that it leads to practical programs and initiatives. He insists that the ecological crisis has a moral character, and defends together the values of respect for life and the integrity of creation. He writes: "Respect for life and the dignity of the human person extends also to the rest of creation, which is called to join [the human] in praising God (cf. Ps. 148:9)."[1]

From 1990 onward, ecological themes appeared often in John Paul II's everyday addresses and his major encyclicals on social justice and moral issues. The term "ecological conversion" was used by John Paul II in an address he gave in 2002. The context is a bleak description of how human beings have disappointed God's expectations, devastating, polluting, and degrading the planet. In such circumstances, John Paul argues, it is essential to support the change of mind and heart at work in many people: "We must encourage and support the 'ecological conversion' which in recent decades has made humanity more sensitive to the catastrophe to which it had been heading."[2] What this conversion might mean is spelled out more fully in the same year in the *Common Declaration of Environmental Ethics* of John Paul II and the ecologically committed Patriarch Bartholomew I. Together they write:

> In our own time we are witnessing a growth of an ecological awareness which needs to be encouraged, so that it will lead to practical programs and initiatives. An awareness of the relationship between God and humankind brings a fuller sense of the importance of the relationship between human beings and the natural environment, which is God's creation and which God entrusted to us to guard with wisdom and love (cf. Gen. 1:28). What is required is an act of repentance on our part and a renewed attempt to view ourselves, one another, and the

---

[1] John Paul II, "Message of His Holiness Pope John Paul II for the Celebration of the World Day of Peace, 1 January 1990." http://www.vatican.va/holy_father /john_paul_ii/messages/peace/documents/hf_jp-ii_mes_19891208_xxiii-world -day-for-peace_en.html. Accessed October 9, 2013.

[2] John Paul II, "General Audience Address, January 17, 2001." http://www .vatican.va/holy_father/john_paul_ii/audiences/2001/documents/hf_jp-ii _aud_20010117_en.html. Accessed October 9, 2013.

world around us within the perspective of the divine design for creation. The problem is not simply economic and technological; it is moral and spiritual. A solution at the economic and technological level can be found only if we undergo, in the most radical way, an inner change of heart, which can lead to a change in lifestyle and of unsustainable patterns of consumption and production. A genuine conversion in Christ will enable us to change the way we think and act. It is not too late. God's world has incredible healing powers. Within a single generation, we could steer the earth toward our children's future. Let that generation start now, with God's help and blessing.[3]

The two leaders speak of a genuine conversion in Christ that will enable us to change the way we think and act. This raises the fundamental question about how ecological conversion is related to conversion in Christ. The *Common Declaration* speaks of a "most radical" inner change of heart, leading to a change in unsustainable patterns of consumption and production. I find this way of describing ecological conversion meaningful, and will attempt to explore it in a more detailed way in what follows. The challenge to a real change of outlook, lifestyles, and models of consumption and production was repeated by Pope Benedict in his major social encyclical *Caritas in Veritate*, and in his World Day of Peace Message of January 1, 2010.

Before beginning to reflect more systematically on this theme of ecological conversion, I will mention Pope Francis's articulation of a closely related idea of "protecting," or being custodians of creation. On March 19, 2013, Francis began his ministry as bishop of Rome with a Eucharist in St. Peter's Square. Celebrating the feast of St. Joseph, he speaks in his homily of Joseph as having the vocation to be protector of Mary and Jesus, and he says of Joseph: "In him, dear friends, we learn how to respond to God's call, readily and willingly, but we also see the core of the Christian vocation, which is Christ!

---

[3] John Paul II and Bartholomew I, "Common Declaration on Environmental Ethics: Common Declaration of John Paul II and the Ecumenical Patriarch His Holiness Bartholomew I, Monday, 10 June 2002." http://www.vatican.va/holy _father/john_paul_ii/speeches/2002/june/documents/hf_jp-ii_spe_20020610 _venice-declaration_en.html. Accessed October 9, 2013.

Let us protect Christ in our lives, so that we can protect others, so that we can protect creation!"[4]

Two things are of particular importance here. First, protecting other human beings and protecting creation go together as one vocation. Second, both are interconnected with protecting Christ. I read this as suggesting that we need to protect the place of Christ in ourselves as well as protect the mission of Christ in the world. We are true to Christ when we protect our brothers and sisters, above all the poorest, and when we protect creation. But as Francis points out, the protection of creation is something to which not just Christians but all humans are called:

> The vocation of being a "protector," however, is not just something involving us Christians alone; it also has a prior dimension which is simply human, involving everyone. It means protecting all creation, the beauty of the created world, as the book of Genesis tells us and as Saint Francis of Assisi showed us. It means respecting each of God's creatures and respecting the environment in which we live. It means protecting people, showing loving concern for each and every person, especially children, the elderly, those in need, who are often the last we think about. It means caring for one another in our families: husbands and wives first protect one another, and then, as parents, they care for their children, and children themselves, in time, protect their parents. It means building sincere friendships in which we protect one another in trust, respect, and goodness. In the end, everything has been entrusted to our protection, and all of us are responsible for it. Be protectors of God's gifts![5]

Francis calls on leaders of peoples:

> Please, I would like to ask all those who have positions of responsibility in economic, political and social life, and all men

---

[4] Homily of Pope Francis, Saint Peter's Square, Tuesday, March 19, 2013. http://www.vatican.va/holy_father/francesco/homilies/2013/documents/papa -francesco_20130319_omelia-inizio-pontificato_en.html. Accessed October 6, 2013.

[5] Ibid.

and women of goodwill: let us be 'protectors' of creation, protectors of God's plan inscribed in nature, protectors of one
another and of the environment.

This protection of creation and of human beings, he says, involves
not only compassion but also tenderness:

> To protect creation, to protect every man and every woman,
> to look upon them with tenderness and love, is to open up a
> horizon of hope; it is to let a shaft of light break through the
> heavy clouds; it is to bring the warmth of hope![6]

Toward the end of his homily, Francis speaks of this protection
as central not only to his own vocation as Bishop of Rome but
also as the vocation of each person: "To protect Jesus with Mary,
to protect the whole of creation, to protect each person, especially
the poorest, to protect ourselves: this is a service that the Bishop of
Rome is called to carry out, yet one to which all of us are called, so
that the star of hope will shine brightly. Let us protect with love all
that God has given us!"[7]

In all of these texts, ecological conversion is understood as something to which the whole global community of humanity is called.
And yet there is a conviction that, for Christians, ecological conversion is deeply connected to their relationship with Christ. What do
John Paul and Bartholomew mean when they speak of "genuine
conversion in Christ" in relationship to ecological conversion? How
might we understand what Francis means when he says "protecting
Christ" involves "protecting creation"? It is these questions that I
will seek to explore in the following two chapters.

[6] Ibid.
[7] Ibid.

Chapter 9

Christian Ecological Conversion

Ensuring the long-term health of Earth's atmosphere, land, rivers and seas, and saving its biodiversity requires nothing less than a transformation of humanity. This transformation is what I am calling ecological conversion. Clearly, it must involve the whole human community in a new level of cooperation and commitment to the good of the planetary community of life. For Christians, this conversion necessarily involves the heart of their faith commitment. How is it interconnected to their relationship with Christ? In order to explore both the more universal call to conversion and the particularly Christian way of participating in this conversion, I will begin with the new sense of planetary spirituality that is shared by people of various religious traditions. Then I will explore a Christian view of ecological conversion as conversion to Jesus Christ that builds on what has been said earlier in this book.

Planetary Spirituality

As I write this chapter, I am sitting at the kitchen table of a little house south of Adelaide, where I often do my theological work. I look out over a rolling landscape toward the Willunga Hills. These hills are folded into one another in beautiful rounded shapes that draw the eye and quieten the spirit. They run right across the horizon down to the sea on my far right. In between my window and the hills is a wide undulating plain. There are waves of late summer golden paddocks, green vineyards, olive plantations, and stands of eucalyptus. Much closer, in the many greens of the indigenous shrubs that make up the front garden, beautiful little black and white birds with gold on their wings, New Holland Honeyeaters, flit between the branches looking for nectar, while magpies scratch around in the ground for grubs.

This place leads me toward stillness. It offers liberation from the busyness of life, from the multiplicities of demands, and the noise that fills so much of contemporary existence. There is an invitation to be quiet before the mystery it mediates. This is an invitation that is all too easy to resist, but to say yes, to dwell even for a short time in the mystery, is to find healing and peace. It is to be open to all that is in this place as a gift. It is to sense the presence of unspeakable Love at the heart of the natural world.

There are many such experiences in our lives, from hiking to working a vegetable garden, to surfing a big wave, to looking up to the stars far from city lights, to contemplating a single flower in its fragility and beauty. It seems safe to say that interrelationship with the natural world has always shaped the aesthetic and religious lives of human beings. If this is so, then the covering of so much of Earth by concrete and asphalt is a cause of deep concern. But the point I want to stress here is simply that we human beings experience grace, the wonder and mystery of God, in the encounter with the world around us, in mountains and deserts, seas and farmlands, in gardens and parks, in birds and animals. It seems this encounter has always had such an effect.

This human experience of the Spirit is fundamental to everything else that I want to say here. Such experiences of grace occurred, of course, long before human beings realized they lived on a planet, and long before they knew that they belonged to a solar system that is part of the Milky Way galaxy in an expanding universe made up of hundreds of billions of galaxies. There is a sense in which a new, contemporary kind of planetary spirituality had its beginning at a precise time—the Apollo 8 mission to the Moon of 1968.

The crew of this mission were the first human beings to completely leave Earth's gravitational field and the first to reach the Moon. In their initial orbits of the Moon, their whole attention was focussed on the Moon and on their instruments. But eventually, as they swung back around from the dark side of the Moon, they were able to attend to Earth coming into view. Astronaut William Anders took a beautiful color picture of Earth rising over the plane of the

Moon's surface. This photograph, called "Earthrise," contributed to a major change in the way we understand ourselves. It became a symbol of that very change.

"Earthrise" shows Earth as a small, blue, green and white planet set against the darkness of interstellar space. Many astronauts have testified not only to their experience of joy and wonder at the beautiful sight of Earth from space but also to a sense of its vulnerability. For some it has been an explicitly religious experience. For many there has been a sense of amazement in the realization that this fragile object in space contains all that they hold precious, all of the diverse forms of life we know, all of human history, all of human wisdom and culture, and all of human love. Through photographs like "Earthrise," the rest of us have been able to share the experience of the astronauts. It has marked a new moment in human cultural history that has shaped us all.

We share a new experience of the natural world, something that was not available in the same way to Plato, Thomas Aquinas, or Isaac Newton. What is new is the sense that we belong to one planetary community of life on this fragile but beautiful blue, green, and white planet that is our home. This has the potential to break down boundaries and barriers of nation, class, race, and religious tradition. We form one global human community. And we are deeply interconnected with the other species of this planet, and with all the systems that support life, the land, the atmosphere, the seas, and the rivers.

Our experience of the natural world is global. We are invited into a new form of planetary spirituality. This spirituality involves a solidarity that challenges our deep-seated tendencies to make "outsiders" into scapegoats and enemies and to engage in ruthless exploitation of other creatures and their habitats. The new picture we have of Earth offers us a common vision of one interrelated community of life, and this can begin to provide a basis for a human response to the crises we face on our planet.

An authentic planetary spirituality, of course, cannot be simply global. It must be local, grounded in place. A genuine ecological commitment involves this place I am in, whether it be a great city or a small farming community, and, through this particular place, with the community of life on the planet. Ecological spirituality involves this bioregion, this river, this remnant of forest, this ecosystem, this cityscape, this garden, this species, this animal, this tree. A genuine planetary spirituality can only be both local and global.

What is required in this kind of global spirituality is not a romantic view of nature, but one that recognizes the otherness of different species, and that stands in awe before the uncontrollable and the unknown in the natural world, from the wilderness we can still experience on our planet, to the age and size of the universe, to the operations of reality at the quantum level. It will need to be a robust spirituality, capable of recognizing the costs of the evolution that has produced the wonderful variety of species of our planet. These costs include the cycles of life and death, the seeming wasteful abundance of some life-forms, the constant competition, the predation, and the extinctions. These costs coexist with the extraordinary cooperation, mutual interrelationships, and the beauty we find in the evolution of life.

At the heart of this kind of planetary spirituality, for many people of various faiths, is the sense that the natural world comes to us as a gift. It is not simply at our disposal, to be abused or squandered at will. For people of diverse faith traditions there is an explicit conviction that what we experience is the gift of a bountiful, generous God. To encounter the world in this way is to know, in some way, the giver of all that is. To receive the world of which we are a part as a gift can lead to a deepening sense that each life-form has its own place and its own value within the one community of life on Earth. This kind of planetary spirituality involves a growing knowledge of our belonging with, and our interdependence with, other species, and with the atmosphere, the seas, and the land. It can result in an enduring commitment to the well-being of the planetary community.

In the land where I belong, Australia, this kind of spirituality, which is both global and local, will necessarily seek to learn from the spirituality of the indigenous peoples who have lived in and with this land for perhaps 50,000 years. Aboriginal Australians from different parts of this land have their own distinct languages and spiritual traditions, but they have some fundamental things in common—including a sense of the sacredness of the land, and an understanding that the identity of human beings involves the life-task of being custodians of the land. This role is often expressed today as the work of "taking care of country" or "looking after country."

Part of embracing a planetary spirituality in this place will involve consciously learning to "look after country" in solidarity with indigenous Australians. It will also involve solidarity with those who love Earth and its creatures from the perspective of Judaism, Islam, Buddhism, Hinduism, and all the other traditions that are at home in this land. From the perspective of Christianity, the whole movement toward a planetary spirituality can be seen, I believe, as something stirred up by the Holy Spirit, the Creator Spirit who breathes life into all creatures. Christianity is one among many religious traditions that seeks to respond to the crisis we face on Earth. But it has its own place and its own contribution to make. Christians are called to participate with others in the challenges facing the Earth community and to do this from the perspective of their own deepest Christian faith. This will involve their Christian ecological conversion. It is the incarnation, I believe, that gives specific shape to Christian ecological conversion.

Ecological Conversion as Conversion to Christ

The proposal of this book is that a fully trinitarian and radically incarnational theology, like that of Athanasius, can form a basis for a contemporary theology of the natural world and a Christian planetary spirituality. While incarnation is at the center of this perspective, it necessarily involves two other key Christian doctrines—creation on the one hand and the bodily resurrection of the crucified Jesus on the other. They go together, I will suggest, like three interrelated panels of a painting, a triptych. Together they form a narrative of God's love for the whole creation, an interconnected story of creation-incarnation-resurrection. I will sketch this triptych briefly, based on earlier chapters of this book, as the theological basis for Christian ecological conversion.

Creation

Why does God create a universe of creatures? This question is answered in the Wisdom of Solomon: "For you love all things that exist, and detest none of the things that you have made. . . . You spare all things for they are yours, O Lord, you who love the living. For your immortal spirit is in all things" (Wis 11:24-26). God creates stars, mountains, seas, eucalyptus trees, dolphins, wallabies, cockatoos, and human beings out of love. Each is a place of God's life-giving presence. In the Spirit, God is deeply interior to each of them, constantly enabling their being out of love. Each of them is loved into being by God. Anthony Kelly writes:

> Everything comes into being out of the sheer abundance of God's generative and creative love. There is a profound truth in speaking of God's creating the universe "out of nothing." When it comes to God's creative act, nothing is first "there," as it were. But we can put that more positively by saying that God creates "out of love." All that is has been loved into being, so that at the heart of all created being is the sheer and unrestricted Love that God is. To that degree, to exist is to have been, and to continue to be, loved into being.[1]

Cosmology tells us of an observable universe that has been expanding and cooling since its origin in a tiny, dense, hot state 13.75 billion years ago. Biology teaches us that all of life on our planet has evolved by processes that include natural selection from its microbial origin about 3.7 billion years ago. Christians can see God as enabling and empowering all the processes described by the sciences through the presence of the life-giving Spirit, the Energy of Love. They can see each creature as called into being by the creative Word of God—"All things came into being through him, and without him not one thing came into being" (John 1:3). This Word of God can be thought of as the Divine Attractor of creation, drawing each cosmic process, each species on Earth, each creature, to its own unique identity, in interrelationship with others. Each becomes, then, a word of the Word, an expression of the dynamic life of the Trinity. Each is precious to God.

[1] Anthony J. Kelly, *God Is Love: The Heart of Christian Faith* (Collegeville, MN: Liturgical Press, 2012), 11–12.

God's creative act is continuous. It occurs not only at the origin of everything, but at every moment that a world of creatures exists and becomes something new. The whole universe of creatures is called into existence by the creative Word of God in the power of the Spirit, not only at the origins of all things, but constantly. The universe and all it contains, all the dynamic processes at work as the universe expands and cools, and every creature on Earth, all exist over an abyss of nothing, held in existence by nothing else than divine love in the relationship of continuous creation.

Love empowers the universe through the indwelling Spirit and the Word, the divine Attractor. Earth and its creatures, its insects, birds and animals, its forests and seas, its habitats and bioregions, all exist because the God of love is closer to them than they are to themselves, and enables their existence, their interaction, and their becoming in the community of creation. As Athanasius teaches us, in the relationship of continuous creation:

- Each creature partakes of the Word of God through the Holy Spirit.

- God, the Source of All, is *immediately* present, not only to the whole creation, but to each single creature, through the Word and in the Spirit.

- Each creature is an expression of the *generativity and fruitfulness* of the life of the Trinity, where the generative Love that is Source and Origin eternally begets the Word and breathes forth the Spirit. The overflowing Fountain eternally gives rise to the River of living water, of which creatures drink in the Spirit.

- This world of creatures exists *within the delight* of the mutual relations of triune life.

The relationship of creation is one by which each creature continually partakes of the generative Love who is the Source of All, through the Word, the divine Attractor drawing all things to their own individuation within the community of creation, and in the Holy Spirit, the indwelling Energy of Love. Each creature in the community of creation springs from the generosity of divine Love, each is precious—"not one of them is forgotten in God's sight" (Luke 12:6).

Incarnation

The divine love for creatures that Christianity celebrates involves far more than creation, although creation is in itself a most wonderful gift, an unspeakable mystery of love. The central panel of the Christian triptych is that of the incarnation. It tells of a God who enters into matter and flesh, uniting it radically with God's self, transforming it from within. Incarnation is not primarily about the birth of the Savior, but about the whole action of God in Christ—the life, death, and resurrection of Jesus that, in the power of the Spirit, transfigures and deifies creaturely existence.

There is, of course, a Christian view of the incarnation that sees it as coming about simply to repair the damage done by human sin. But there is a deeper view, associated with Maximus the Confessor (580–662) in the East, and with the Franciscans, particularly Duns Scotus (1266–1308), in the West, and taken up more recently by Karl Rahner among many others, that sees God's creation of a world of creatures as always directed toward the incarnation. The incarnation is certainly an unthinkable act of divine mercy and forgiveness. But on this deeper theological view, the incarnation is understood as the very reason for the creation. This means that even in a world without sin, the Word of God would have come to us. Of course, the incarnation is not something that could ever be predicted on the basis of creation. It is a completely unpredictable and radically gratuitous act of divine love. But dwelling within the perspective of the incarnation, given to us only in the Christ event, we can see that God always intended what Pope Francis has called the "revolution of tenderness" that is the incarnation.[2]

Creation, then, is always about God's self-bestowal to creatures in love. God creates a world of creatures in order to give God's very

---

[2] Pope Francis, "Address to CELAM leadership," http://en.radiovaticana.va /news/2013/07/28/pope_francis:_address_to_celam_leadership_/en1-714819. Accessed October 6, 2013.

self to them in love. As Rahner puts it, creation and incarnation are united as distinct aspects of the one process of God's one act of self-bestowal to creatures.[3] This echoes the late Pauline texts that speak of God's plan for the fullness of time to "gather up" the whole of creation in Christ (Eph 1:10), and to "reconcile" the whole creation in Christ and bring the whole of reality to its shalom in God (Col 1:20). It reflects the opening hymn of John's gospel, where we hear how God created the whole universe of creatures through the eternal Word (John 1:3), and how this Word of creation became flesh (John 1:14).

In the incarnation, God unites the whole of creaturely reality with God's self from within, thus bringing about its deification—its unforeseeable transfiguration and fulfillment in the Trinity. The divine union with the creaturely humanity of Jesus involves not just his humanity, and not just the rest of the human community, but all living things and all matter. It involves all the processes of cosmic and biological evolution. God becomes a creature of flesh and blood, made of atoms that are produced in stars, shaped by evolutionary history, subject to pain and death, in solidarity with the whole community of life on Earth. Incarnation, then, is both "deep" and "wide" in its effects on the world of creatures. As Pope John Paul II has written: "The incarnation of God the Son signifies the taking up into unity with God not only of human nature, but in this human nature, in a sense, of everything that is 'flesh': the whole of humanity, the entire visible and material world."[4] Incarnation involves all the other living creatures of our planet and all that supports their life.

Incarnation is certainly a most precious event of radical forgiveness for humankind. And, as Athanasius tells us, it is an act of divine self-revelation—the Wisdom of God already manifest in the

---

[3] Karl Rahner, *Foundations of Christian Faith: An Introduction to the Idea of Christianity* (New York: Seabury Press, 1978), 197.

[4] John Paul II, *On the Holy Spirit in the Life of the Church and the World*, par. 50. http://www.vatican.va/holy_father/john_paul_ii/encyclicals/documents/hf_jp-ii_enc_18051986_dominum-et-vivificantem_en.html. Accessed October 9, 2013.

diversity of creation all around us now comes to us and meets us in our own humanity in the midst of biological life. But incarnation is more than forgiveness, more than revelation. It is a divine act that transforms the whole creation. It is the beginning of the end of death. God becomes a creature of matter and flesh in order that human beings and with them the rest of creation might be deified and transformed in God, finding their creaturely fulfillment by participating in the life of the Trinity. This process has begun in our world through the resurrection of the crucified Christ, the beginning of new creation at work in our world.

The Resurrection of the Crucified Jesus

An incarnational spirituality culminates in what is depicted in the third panel of the triptych, the bodily resurrection of the crucified Jesus. Jesus transfigured in glory is the promise and the beginning of the transfiguration of humanity and, with them, of the wider creation of which they are a part. Paul tells us that this transfiguration in Christ has already begun in us through the Spirit: "And all of us, with unveiled faces, seeing the glory of the Lord as though reflected in a mirror, are being transformed into the same image from one degree of glory to another; for this comes from the Lord, the Spirit" (2 Cor 3:18). In another context, he speaks of the further transformation that is our participation in resurrection, when the risen Christ will transform our humble bodies so that they are conformed to his glorious body (Phil 3:21).

Christians have sometimes thought of themselves as being saved by being taken out of this world. The whole argument of this book has been that such a view is unfaithful to the central Christian doctrines of creation, incarnation, and resurrection. We are saved not by being taken out of the rest of creation, but by the transformation of ourselves and the rest of the creation in Christ. As Paul describes things: "The creation itself will be set free from its bondage to decay and will obtain the freedom of the glory of the children of God. We know that the whole creation has been groaning in labor pains until now; and not only the creation, but we ourselves, who have the first fruits of the Spirit, groan inwardly while we wait for our adop-

tion, the redemption of our bodies" (Rom 8:21-23). Biblical scholar
N. T. Wright sees this as the "greatest Pauline picture of the future
world," and the "deepest New Testament answer to the problem
of evil." At the heart of it is the promise that God will do for the
whole universe what God did for Jesus at Easter.[5]

Earlier I referred to Rahner's conviction that a truly Christian
understanding of God is reached only when we can think of a God
who is radically distinct from the creation but gives God's very self
to creation as its own fulfillment in the incarnation of the Word.
In such an incarnational view, Rahner says, God is understood as
"the very core of the world's reality and the world is truly the fate
of God."[6] In the Word made flesh, matter and flesh are taken to
God's self in a radical way. In the resurrection and ascension of the
crucified Jesus, matter and flesh are forever in God. Rahner insists
that what is central to Christianity is that in the incarnation, and its
culmination in resurrection, God commits God's self to this world,
to this Earth and its creatures and does so eternally.

I also referred earlier to Torrance's words: "God has decisively
bound himself to the created universe and the created universe to
himself, with such an unbreakable bond that the Christian hope of
redemption and recreation extends not just to human beings but to
the universe as a whole."[7] In the risen Christ, part of the biological
community of Earth is forever transfigured in God, the promise and
the beginning of the transformation of all things in God. In the risen
Christ we have a divine promise that "all things" will be transformed
(Rom 8:18-25; Col 1:15-20; Eph 1:10; Rev 5:13-14). We can hope
that it includes in some way every wallaby, dog, and dolphin, and the
whole of evolutionary history. Creatures will find their fulfillment
in God according to their own identity and receptivity. We have no
good imaginative picture or clear conceptual knowledge of what the
future of creation might be in God. All we have is the promise of

---

[5] N. T. Wright, *Evil and the Justice of God* (London: SPCK, 2006), 75.

[6] Karl Rahner, "The Specific Character of the Christian Concept of God,"
Theological Investigations 21 (New York: Crossroad, 1988), 191.

[7] Thomas Torrance, *The Christian Doctrine of God: One Being Three Persons*
(Edinburgh: T & T Clark, 1996), 244.

God in Christ, that what has happened in Jesus' resurrection is the beginning of the fulfillment of the universe in God.

In pondering the triptych of creation, incarnation, and the resurrection of the crucified Jesus, we come to know that we cannot love God without loving God's beloved creatures. We cannot follow Jesus, the Word made flesh, without embracing the matter and flesh embraced in his incarnation. The three Christian doctrines form the basis for a Christian commitment to this Earth and its creatures. Conversion to Christ involves conversion to love of the neighbor; it involves breaking all barriers of insiders and outsiders in love of the enemy; it involves a love for this Earth and all its creatures, in ecological conversion.

I have been proposing that to be converted authentically to love of Jesus as the Word of God made flesh, and to following his way as a disciple, involves being converted to love for our planetary community of life—the creation eternally embraced by God through the Word and in the Holy Spirit, in the divine acts of creation, incarnation, and resurrection. There is a particular urgency to this truth for us today, as we face the crisis of life on our planet, but it has always been true, as saints, like Cuthbert, Hildegard of Bingen, Francis of Assisi, John of the Cross, and many others, have taught us.

Conversion to Christ today involves ecological conversion, and the deep reason for this is given in the triptych of creation, incarnation and resurrection. This, I believe, is the theological foundation for what Pope Francis says in his inauguration homily, summed up in one of his tweets, "Let us keep a place for Christ in our hearts, let us care for one another, let us be loving custodians of creation."[8] Caring for one another, above all the poorest and most needy, and becoming more and more a loving custodian of creation—this is to keep a place for Christ in our minds, hearts, and actions. Being converted more deeply to Christ involves deepening ecological conversion.

---

[8] Pope Francis, "Let us keep a place for Christ in our hearts, let us care for one another, let us be loving custodians of creation," March 19, 2013, 3:55 a.m., tweet. https://twitter.com/Pontifex/status/313966951368105985.

Chapter 10

The Community of Creation

What does ecological conversion mean for our concept of ourselves in relationship to the rest of creation? In attempting some response to this question, I will outline a biblical view of humans as not only made in the image of God but also as called to cosmic humility. Then I will propose that both of these aspects of biblical theology are embraced within the biblical picture of humans as participants with other creatures in the one community of creation before God.[1] In the second section of this chapter, I will offer some reflection on five dimensions of human existence that are capable of being transformed by the grace of ecological conversion. I will conclude with a very brief section on ecological conversion as a way of praying.

Humans and Other Creatures

In 1967 medieval historian Lynn White published an article in the journal *Science* entitled "The Historical Roots of our Ecological Crisis."[2] In White's analysis, Christianity's biblical faith embodies a linear notion of time, which implies faith in continual progress and encourages the concept of endless growth. Furthermore, he says,

---

[1] This is a development of material discussed in Denis Edwards, *Jesus and the Natural World* (Mulgrave, Vic.: Garratt Publishing, 2012), and in "Anthropocentrism and its Ecological Critique: A Theological Response," in *Being Human: Groundwork for a Theological Anthropology for the 21st Century*, ed. David Kirchhoffer, et al. (Preston, Vic.: Mosaic Press, 2012), 107–21.

[2] Lynn White, "The Historical Roots of Our Ecological Crisis," *Science* 155 (March 10, 1967): 1203–07. My references to this article come from *This Sacred Earth: Religion, Nature, Environment*, ed. Roger Gottlieb (New York: Routledge, 1996), 184–93.

Christian faith rests on a creation account in which human beings are made in the divine image and the rest of creation is made to serve the human. White finds Western Christianity anthropocentric in the extreme:

> Especially in its Western form, Christianity is the most anthropocentric religion the world has seen. As early as the second century both Tertullian and St. Irenaeus of Lyons were insisting that when God shaped Adam he was foreshadowing the image of the incarnate Christ, the Second Adam. Man shares, in great measure, God's transcendence of nature. Christianity, in absolute contrast to ancient paganism and Asia's religions (except, perhaps, Zoroastrianism), not only established a dualism of man and nature but also insisted that it is God's will that man exploit nature for his proper ends.[3]

White sees Christianity as demystifying nature, and thus allowing it to be exploited with indifference. He finds that Christian interest in natural theology supports the scientific exploration of nature and the eventual dominance of science and technology.[4] When combined with Christianity's conviction of the rightness of human mastery over nature, factors such as these have led to the current ecological crisis. White thus finds that "Christianity bears a huge burden of guilt"[5] for the fact that human exploitation of the natural world is out of control.

Many have questioned aspects of White's historical argument. Ernst Conradie has tracked the discussion and provides a helpful bibliography.[6] Peter Harrison, for example, accepts that some biblical texts have been used as resources for Western exploitation, but denies that this played a major role before the rise of modern science in the seventeenth century.[7] Certainly there are key figures

[3] White, "Historical Roots," in Gottlieb, *This Sacred Earth*, 189.

[4] Ibid.

[5] Ibid., 191.

[6] Ernst Conradie, *Christianity and Ecological Theology: Resources for Further Research* (Stellenbosch, South Africa: Sun Press, 2006), 61–67.

[7] Peter Harrison, "Having Dominion: Genesis and the Mastery of Nature," in *Environmental Stewardship: Critical Perspectives, Past and Present*, ed. R. J. Berry (London: T & T Clark, 2006), 17–31.

like Francis Bacon (1561–1626) for whom scientific progress is the implementation of the God-given mandate of human dominion over creation found in Genesis 1:28. In Bacon's view, this dominion, which is partially lost by human sin, is to be restored in the great enterprise of scientific achievement.[8]

While the "dominion" text in Genesis has undoubtedly been invoked to justify exploitation, I am not convinced that the biblical and Christian tradition is simply, or necessarily, extremely anthropocentric. Unlike so much of the theology of recent centuries, pre-Reformation and pre-Enlightenment Christianity, in its biblical, patristic, and medieval expressions, always revolved around not only God and humanity but also the rest of natural world. There are instances when Christian thinkers have taken clearly anthropocentric positions, as when Aquinas sees animals and plants as existing only for the use of human beings.[9] Such views coexisted, however, in Aquinas and others, with a broad theology of God's good creation, and alongside the Benedictine theology of care for creation, and the long tradition of saints like Cuthbert and Francis treating animals as fellow creatures before God.

Thomas Berry, a cultural historian and prophetic figure in the Christian ecological movement, has been a strong critic of anthropocentrism in the Christian tradition. To save Earth, Berry argues, we need consciously to shift from an anthropocentric to a biocentric norm of reference.[10] Human beings must come to see themselves as part of the community of life on the planet and this will involve "the change from an exploitative anthropocentrism to a participative biocentrism."[11] He sees humans as needing to learn to go beyond their cultural coding, shaped by anthropocentric traditions that determine language, intellectual insights, educational programs, spiritual ideals, imaginative power, and emotional sensitivities. They

---

[8] On this history see Richard Bauckham, *Living with Other Creatures: Green Exegesis and Theology* (Waco, TX: Baylor University Press, 2011), 14–62.

[9] Thomas Aquinas, *Summa Theologiae*, II-II, 64, 1.

[10] Thomas Berry, *The Dream of the Earth* (San Francisco: Sierra Club Books, 1988), 21, 30, 165–66, 202–10.

[11] Ibid., 169.

need to go deeper, he argues, to live from their genetic coding, where they are an integral part of the universe and the community of life on Earth.

As much as I respect Berry's prophetic witness, I am not convinced that the remedy for human arrogance is to move from an anthropocentric approach to a biocentric one. Theologically, I think that what is needed is neither the extreme anthropocentrism that offers no respect for the dignity of other creatures, nor the biocentrism that would seem to reject the unique dignity of the human person, but a position more nuanced than either anthropocentrism or biocentrism—one that is explicitly theocentric. The biblical/theological position is not necessarily anthropocentric and it is not simply biocentric. It sees human beings and with them the whole creation in relation to God. It contains the resources for a broader and more inclusive understanding of the relation between the human being and the rest of God's creation. Such a theocentric view, I hope to show, can nevertheless be at one with Berry in his vision of human beings as part of the one community of creation before God.

Building directly on insights offered by biblical scholar Richard Bauckham,[12] I will outline a constructive theology of the human in relation to the wider creation as involving three interconnected and complementary dimensions of the biblical/theological tradition: humans as uniquely called to serve and protect the wider creation, humans as challenged to cosmic humility before God and God's creation, and humans as participants in the one community of creation on Earth before God.

Called to Serve and Protect Creation

In opposing exploitative anthropocentrism, I see little sense in attempting to minimize the uniqueness of the human. From a scientific perspective it is clear that something extraordinary happens with the emergence of the human—the human brain is by far the most complex thing we know in the universe. From an ethical perspective,

---

[12] Richard Bauckham, *Bible and Ecology: Rediscovering the Community of Creation* (London: Darton, Longman & Todd, 2010).

it is fundamental to require of humans that they take responsibility for the damage they have done to the Earth and its other creatures and begin to act for Earth's healing. From a theological perspective, it is clear that Christianity's commitment to the incarnation involves a unique view of the human vocation, even as it also demands a clear-eyed view of human sin.

Unfortunately, Christianity's concept of the human is often simply identified with an uncritical reading of Genesis 1:26-28, in which human beings are commanded to have "dominion" over other creatures and "subdue" the Earth. "Dominion" can all too easily be understood as domination and exploitation. "Subduing" is extremely harsh language that needs to be interpreted as time conditioned and as dangerously inappropriate when used unthinkingly in today's ecological context. Above all, these words need to be understood in their context in the book of Genesis, where God celebrates the goodness of the whole creation and blesses its fruitfulness, and in the context of the wider biblical view of God's love and care for all creatures.

I hold to the enduring importance of the idea found in this same text of Genesis that human beings are made in the image of God. Unlike my colleague Norman Habel, I think that the idea of being made in the image of God can be freed from its connection with the language of subduing.[13] With Claus Westermann, I take it to mean that humans are creatures with whom God can speak (Gen 1:28-30), creatures to whom God relates interpersonally.[14] I adopt the theological view that God can be thought of as relating to all creatures in terms of their own proper nature and integrity. I see God's relationship with humans in terms of their interpersonal nature. In this context, the teaching that humans are made in the image of God can mean that human beings are called to share in the divine love and respect for other creatures. They are to relate to them as God does, with something like God's feeling for them. Humans, in the divine image, have a responsibility for creation as humble servants of God.

[13] Norman Habel, *An Inconvenient Text: Is a Green Reading of the Bible Possible?* (Adelaide: ATF Press, 2009), 1–10.

[14] Claus Westermann, *Creation* (Philadelphia: Fortress Press, 1974), 58.

The relationship of humans to other creatures receives positive expression in the second creation account in Genesis, where God takes the newly created human to the Garden of Eden in order "to till it and keep it" (2:15). Habel points out that the Hebrew word usually translated as "till" (*abad*) has the basic meaning of "serve." Human beings are thus called to serve and preserve the land, and in doing so they contribute to the greening and completion of creation. Habel notes that in the gospels, we find Jesus rejecting all domination and replacing it with loving service of others (Mk 10:42-45). He rightly proposes that we need to apply this not only to other humans but also to the other creatures that make up the community of life.[15]

Earlier, I discussed the homily of Pope Francis in the Eucharist that began his ministry as bishop of Rome, where he spoke of the human vocation to be a protector or custodian of the whole creation, which involves respecting each of God's creatures and each of their habitats. What Francis calls the human vocation to be a protector of the natural world is closely related to the long tradition, particularly strong in Protestant ecological theology, of speaking of the human vocation to be faithful "stewards" of creation.[16] This theology of stewardship supports a great deal of ecological commitment. It is important, I believe, to recognize that the language of stewardship, like the language I am using—the call to serve and protect the creation—is part of the picture in ecological theology but not the whole. To make it the whole would be to slip again into distorted anthropocentrism that fails to recognize God's direct relationship with other creatures that is not mediated by the human. Stewardship or custodianship can be rightly understood as one important aspect of ecological theology, the aspect focused on human responsibility and participation. But it needs to be seen within the larger biblical context of the call to cosmic humility, and the fuller biblical picture of human beings as part of the one community of creation.

---

[15] Habel, *An Inconvenient Text*, 68–77.

[16] For various Christian views on the stewardship of creation, see *Environmental Stewardship: Critical Perspectives—Past and Present*, ed. R. J. Berry (New York: T & T Clark International, 2006).

Called to Cosmic Humility

Alongside the affirmation of the human made in the image of God and called to serve and protect creation, there is a biblical/theological tradition that puts humans in their place and challenges them to cosmic humility, not only before God, but also before God's other creatures. In this tradition, these creatures are seen not as dependent on human beings but as having their own unique relationship to the living God. I will point briefly to the way these ideas appear in two of the major biblical treatments of creation, the book of Job and Psalm 104.

In Job we find the longest passage in the Bible about nonhuman creation. It comes after thirty-five chapters in which Job's friends offer unsatisfactory explanations for the evil that has befallen Job, and Job accuses God of being unjust. Finally, at the beginning of chapter 38, God answers Job from the whirlwind. The divine response is not an explanation, but a challenge to Job to look at the universe that God has created. God invites Job "into a vast panorama of the cosmos, taking Job on a sort of imaginative tour of his creation, all the time buffeting Job with questions."[17] God's questions challenge Job's worldview and call him to a new way of seeing. The first question immediately puts Job in his place and sets the tone:

> Where were you when I laid the foundation of the earth?
> Tell me if you have understanding.
> Who determined its measurements—surely you know!
> Or who stretched the line upon it?
> On what were its bases sunk, or who laid its cornerstone
> when the morning stars sang together
> and all the heavenly beings shouted for joy? (Job 38:4-7)

God then lays out before Job the immensity and wonderful order of the physical universe (38:8-38). Job is challenged to a take a humble stance before God's creation of the land, the sky, the sea, the snow, the rain, the clouds, and the constellations of the night sky. God then turns from the physical universe to the biological, ask-

---

[17] Bauckham, *Bible and Ecology*, 39.

ing Job: "Can you hunt the prey for the lion, or satisfy the appetite of the young lions?" God puts before Job the raven, the mountain goat, the deer, the wild ass, the wild ox, the ostrich, the horse, the hawk, and the eagle. Each of the ten animals or birds mentioned is wild, independent of human beings. Each stands in relationship to God in its own right. There is no human mediation. Each creature is described in detail, in species-specific ways. Job is challenged not only to humility but also, it seems, to share God's joy in them, in all their distinctive otherness from the human.

The same interest in the unique and independent relationship of other species with the Creator, and the same invitation to humans to share God's delight in their diversity, is found in the great biblical song of creation, Psalm 104.

> The trees of the LORD are watered abundantly,
> the cedars of Lebanon that he planted.
> In them the birds build their nests;
> the stork has its home in the fir trees.
> The high mountains are for the wild goats;
> the rocks are a refuge for the coneys.
> You have made the moon to mark the seasons;
> the sun knows its time for setting.
> You make darkness, and it is night,
> when all the animals of the forest come creeping out.
> The young lions roar for their prey,
> seeking their food from God.
> When the sun rises, they withdraw
> and lie down in their dens.
> People go out to their work
> and to their labor until the evening.
> O LORD, how manifold are your works!
> In wisdom you have made them all;
> the earth is full of your creatures. (Ps 104:16-24)

Human beings are fellow creatures with other animals before God. All are related among the "manifold works" of God. All are made in divine Wisdom. God breathes into each the breath of life (v. 30). Reading and pondering Job and Psalm 104 in today's world can lead us to a humbler stance before the mystery of the enormous age and size of the observable universe, the quantum depths of physical reality, and the extraordinary story of the evolution of life. We are invited into a deeper respect for other species in their distinctive difference from our own, as having their own direct relationship to God their Creator, and as having their own God-given integrity. These texts not only bring human beings to cosmic humility but also point to the place of human beings within the one community of creation before God.

The Community of Creation

In the rich biblical tradition, humans are seen as united with other creatures in one absolutely fundamental characteristic—they are God's creatures. Along with all other entities of our universe, we are fellow creatures before the one God who is Creator of all. Biblical faith not only leads us to recognize our common creaturehood but also calls us to see ourselves as forming a community of praise of God the Creator. This theme appears in many of the psalms. Psalm 148 is a beautiful example:

> Praise him, sun and moon;
> praise him, all you shining stars!
> Praise him, you highest heavens,
> and you waters above the heavens!
> Let them praise the name of the LORD,
> for he commanded and they were created.
> He established them forever and ever;
> he fixed their bounds, which cannot be passed.
> Praise the LORD from the earth,
> you sea monsters and all deeps,
> fire and hail, snow and frost,
> stormy wind fulfilling his command!
> Mountains and all hills,
> fruit trees and all cedars!
> Wild animals and all cattle,
> creeping things and flying birds!
> Kings of the earth and all peoples,
> princes and all rulers of the earth!
> Young men and women alike,
> old and young together! (Ps 148:3-12)

More than thirty categories of creatures are addressed in this psalm.[18] All are seen as praising God in their own unique way. This same theme is found in the *Song of the Three Young Men* from Daniel, sung regularly in the Morning Prayer of the church on Sundays and feast days. The whole world of creatures is called to praise and bless the Creator: "All that grows in the ground . . . seas and rivers . . . you springs . . . you whales and all that swim in the waters . . . all birds of the air . . . all wild animals and cattle . . . all people on earth" (Dan 3:79-81). One of the most beautiful and appealing developments of this theological vision is found in Francis of Assisi's *Canticle*, where Francis addresses other creatures as brother and sister and deliberately popularizes a theology of the kinship of human beings and other creatures before God.[19]

In every Eucharist we form a community of creation united in praising God. When the Christian community brings gifts of bread and wine to the altar, it symbolically brings with these gifts the whole creation. The great prayer over the gifts is a prayer of thanksgiving for God's act of creation as well as for God's saving action in Christ. It is an act of praise of God from the whole community of creation. These themes are explicit in several of the Catholic Eucharistic prayers. Immediately after the *Sanctus*, in the Eucharistic Prayer III (often used in parishes on Sundays), the gathered community formally praises God in union with all other creatures: "You are indeed Holy, O Lord, and all you have created rightly gives you praise." In the Preface of Eucharistic Prayer IV we find: "With them we, too, confess your name in exultation, giving voice to every creature under heaven." The gathered community gives vocal praise in union with all of God's creatures, with flowers, trees, parrots, and kangaroos, who constantly give God praise by being what they are. To think of a great Sequoia tree, or a laughing kookaburra, or a dolphin riding the surf as praising God is to enter a biblical way of thought in which we see ourselves as fellow creatures with others in

---

[18] Ibid., 77.

[19] For a fine treatment of this whole Franciscan tradition, see Dawn M. Nothwehr, *Ecological Footprints: An Essential Franciscan Guide for Faith and Sustainable Living* (Collegeville, MN: Liturgical Press, 2012).

the one communion of creation, all of which is to be transformed in Christ (Rom 8:18-25).

Ecological Conversion as Transformation

Can more be said about this ecological conversion which is our human and Christian vocation? I will suggest that it is a grace that involves a transformation in five aspects of human existence. This transformation leads to a new way of thinking, seeing, feeling, acting and living, each mutually interrelated with all of the others. For the Christian believer, each involves a deepening of conversion in Christ, the divine Attractor of the whole creation and the companion of our personal journeys. Each aspect of this conversion, insofar as it is genuine, is the gift of the Spirit in us, the Energy of Love leading us into the new.

Conversion: A New Way of Thinking

In describing ecological conversion as a "new" way of thinking, seeing, feeling, acting, and living, I mean to suggest that the Spirit of God is calling us to something new in our present context of planetary crisis. This "new" will involve, in part, recovering and receiving anew what has been given long ago. It is based upon the Scriptures and the great Christian tradition. But it is appropriated in a new way in response to a new context. The Spirit is calling us to a new reception of the Gospel of God given to us.

Conversion of our thinking involves a turning from a dangerously anthropocentric way of thought: one that sees other creatures as existing only for human use and exploitation, that understands human beings in individualistic terms, that exults the spirit at the expense of matter and the body, and that sees salvation as being taken out of creation. This book has been an attempt to describe the way of thinking to which we are called. I have been summing this up in what I have called the Christian triptych: God's immediate presence through the Word and in the Spirit in continuous creation that enables each creature to exist and to interact in a community of creation, and that enables the creation to have its own capacity to evolve into the new; God's self-giving to creation in the incarnation of the Word and the outpouring of the Spirit, by which humans are adopted and deified and the whole creation is transformed from within; God's resurrection of the crucified Jesus, which means that, in the risen Christ, matter and flesh are forever in God, and which constitutes an unbreakable promise of God that with human beings the whole creation will reach its proper fulfillment.

This triptych is not new—it is celebrated in every Easter Triduum and every Sunday Eucharist. What is new is its application to the way we think and act with regard to our fragile planet and all its inhabitants today. It has three consequences that can be seen as central to a process of ecological conversion. The first is that, without in any way diminishing our commitment to the unique dignity and value of the human person made in the image of God, we come to recognize the uniqueness and dignity of other species, as words of the Word, and come to see with Francis of Assisi that we are fellow creatures before God. In this sense we are sisters and brothers to them and they are kin to us. We are part of the one family of creation.

A second consequence is that in this kinship of creation, each species and each creature has its own value before God. Other creatures do not get their value because of their usefulness to human beings, as the great theologian Thomas Aquinas taught. On this issue I believe we need to depart from Thomas for a deeply theological reason—God's love and respect for each creature. We cannot think of the relationship between God and other creatures only as mediated through human beings. Other creatures have their own relationship with the Creator. It is this unique relationship that allows them to exist. Each of them partakes of the Word in the Holy Spirit. Each of them is promised its own unforeseeable future and fulfillment in God. These creatures have their own relationship with God, their own integrity, and their own God-given intrinsic value.

Closely related to the ideas of the kinship of creation and the value and integrity of other creatures is a third: we form one community of life on Earth. Ecological conversion involves a way of thinking about ourselves as interdependent and interrelational. It involves moving from an individualistic conception of the human person to seeing humans as constituted by relationships with other human beings, with the natural world of which they are a part, and with the triune God. Undergoing conversion in our thinking means coming to understand ourselves as one community of life on this fragile, beautiful, blue, green, and white planet, and as responsible for its future.

A New Way of Seeing

Closely related to the way we think is the way we see things. What we think determines how we see and what we allow ourselves to see determines the way we think. Jesus said: "The eye is the lamp of the body. So, if your eye is healthy, your whole body will be full of light; but if your eye is unhealthy, your whole body will be full of darkness. If then the light in you is darkness, how great is the darkness!" (Matt 6:22-23). How we see things determines our stance before the reality around us. And how we see things depends upon how we have learned to see, what we allow ourselves to see, how we have trained ourselves to see.

Seeing aright involves a conversion in our seeing. It involves discipline and virtue. The ecological theologian Sallie McFague, drawing on an insight from feminist theology, distinguishes between two ways of seeing. She calls them the loving eye and the arrogant eye. The loving eye is a way of seeing the other that really attends to the other, that allows this other to be truly different from oneself and one's own presuppositions, and that attempts to respect the integrity of the other. The arrogant eye sees the other as useful to one's own purposes. It objectifies, manipulates, uses, and exploits.[20]

The loving eye requires a kind of detachment, to really see the other rather than simply the projection of one's self or one's own needs. This, of course, is true of human relationships, but it is also true of the way we see the natural world. It is when our seeing has an element of detachment that our eye can be truly loving, based on the reality of the other rather than on our own needs and fantasies.

---

[20] Sallie McFague, *Super, Natural Christians: How We Should Love Nature* (Minneapolis, MN: Fortress Press, 1997), 30–36. McFague refers to the work of Marilyn Frye, *The Politics of Reality: Essays in Feminist Theory* (Trumansburg, NY: Crossing Press, 1983), 53–83.

Seeing with a loving eye requires seeing the natural world around us with attention to detail. The model for this kind of seeing is the scientific naturalist who closely observes insects, animals, or birds, noticing particulars and differences, knowing species by name. Knowing the scientific names is not the point. The point is close observation. The loving eye is not about attempting to comprehend totally or to control, but rather accepting the mystery of the other in humility. Conversion in our seeing of the natural world means moving from a looking that is exploitative or that ignores the true reality of the other to a mature, humble, loving attentiveness to all its specificity and otherness.

A New Way of Feeling

Undergoing conversion in thinking and seeing cannot be separated from transformation in the way we feel. Ecological conversion involves a deepening of our feeling for the natural world of which we are a part, and a deepening of the feeling of belonging to the wider community of life on Earth. Learning to see with a loving eye can lead to feelings of wonder and gratitude. When these feelings arise in us, we can respond either by quickly moving on or by learning to stay with the experience of wonder and gratitude.

Ecological conversion means learning to receive and dwell in such feelings, whether they come from observing stars at night or seeing and smelling a garden bed before me. It can sometimes mean allowing myself to feel my interconnectedness with the stars and the galaxies and all that makes up the long, slow 3.7-billion-year history of the evolution of life on Earth.

In a fundamental way, ecological conversion involves a continuing deepening in the capacity to feel with other creatures, for empathy with them. This does not mean adopting a sentimental or romantic view of the natural world. We need a robust view of the complexity of the natural world, with all its evolutionary patterns. But it does involve a feeling for the pain of other creatures, and for the frustration of their natures in, for example, the factory farming of animals, a feeling that can then give rise to ethical conviction and action.

Kinship with the other species of our planet is not simply an intellectual conviction, but something that is felt. The experience of feeling that we belong with other species and other creatures is something for which we can make space, and to which we can attend. This, then, can become part of our prayer. We learn to sing praise to God with all other creatures, with "Mountains and all hills, fruit trees and all cedars! Wild animals and all cattle, creeping things and flying birds" (Ps 148:9-10). Praising God with other creatures deepens our feeling of kinship with them. Feeling our kinship with other creatures, feeling part of the community of creation, we begin the great eucharistic prayer with the words: "All you have created rightly gives you praise."

A New Way of Acting

Thinking, seeing, and feeling shape our action, and action shapes the way we think, see, and feel. Action is intrinsic to ecological conversion. Ecological action can begin in very small changes in the way we act in the home, workplace, school, church, local neighborhood, or political arena. If such actions are not to lead to dead ends, they need to be followed by reflection that leads to further action. Actions that seem small can then be transformative and lead to deeper conversion and involvement with others in common projects and political action.

Action for justice for human beings and ecological action go together. Conversion to Christ involves both. They are not to be played against one another. Christian commitment to the community of life on our planet will involve commitment to the liberation of human beings from poverty and oppression and commitment to the well-being of all the other species of Earth. The two very often intersect in obvious ways: protecting the waters of our rivers, lakes, and aquifers is a basic ecological issue; it is also a central and extremely urgent matter of justice, where millions of human beings already have no access to clean drinking water. As Pope Francis puts it, we are called to be protectors of creation and protectors of one another, above all, the poorest. These are two essential dimensions of our one God-given human vocation.

One of the insights offered by liberation theologians who have long been committed to the poor is that truly Christian political action needs to be grounded in prayer, in the mystical.[21] What is true of commitment to justice for the poor of the Earth is also true of ecological commitment. The political and the mystical need to go together as dimensions of ecological conversion. It is only the mystical that can enable us to hope against hope, to continue to act with integrity and love in the political and the personal spheres in times of adversity and failure, up to and including death.

Those committed to a new way of being on Earth discover not only the joy of God's presence in the beauty of the natural world but also find that they have the same need for discipline and perseverance in the dark times, as is the case in traditional forms of mysticism. Commitment to the good of the Earth community can involve joy and solidarity, but it can also involve something like the dark night of St. John of the Cross. In a Christian vision, prayer and lifelong ecological commitment belong together. Edward Schillebeeckx has said: "Without prayer or mysticism politics soon becomes cruel and barbaric. Without political love, prayer or mysticism soon becomes sentimental or uncommitted interiority."[22]

A New Way of Living

Ecological conversion involves not only particular actions but also the action that is one's whole life. It involves me living the vocation that is my own, with a loving eye for the natural world around me, committing myself to being an Earth-loving and ecologically acting parent, homemaker, farmer, teacher, builder, business person, church leader, or student. It is a matter of being called to a lifelong commitment to the good of the Earth community, in and through all that makes up the reality of my daily life, and seeing this as the way of following Jesus the Wisdom of God in this time.

---

[21] See, for example, Gustavo Gutiérrez, *We Drink from Our Own Wells: The Spiritual Journey of a People* (Maryknoll, NY: Orbis, 1984, 2006); *The Mystical and Political Dimension of Christian Faith*, ed. Claus Geffré and Gustavo Gutiérrez, Concilium 96 (New York: Herder and Herder, 1974).

[22] Edward Schillebeeckx, *Jesus in Our Western Culture: Mysticism, Ethics, and Politics* (London: SCM Press, 1987), 75.

Jesus tells us: "For where your treasure is, there your heart will be also" (Luke 12:34). Conversion to the way of Jesus involves finding liberation from the patterns of individualism, consumption, and possession that dominate our society. If we examine our way of life, we can find that we can act as if certain things really matter. But these do not necessarily represent our deepest values. What is it that really matters? Is it owning more that really matters to me? Is it trivial entertainment? Is it constant cell phone calls? Is it appearing successful? Is it being busy or seemingly irreplaceable that matters most to me in life? The words of Jesus offer a liberating challenge: What is it that we really do treasure? What is it that really matters?

All of us, I think, would hope we would be provided with the basics of shelter, clothing, food, drink, education, and reasonable medical care. All who are Christian and many who are not would want to give priority to their relationship with the living God. But granting these things, what is it that we would see as the true goods of life? I think answering this question in some way is central to ecological conversion.[23] We would all differ somewhat in our answers. Some things that occur to me are these:

- Loving relationships, family and friends
- Contributing to community life
- Birds, animals, flowers, trees, deserts, forests, and beaches
- Meaningful work of some kind
- Creativity
- Continual learning
- Music
- An imaginative life
- Shared meals

Our own personal list tells us what matters to us. This is what we value, where our treasure truly is, where our heart really is. Ecological conversion involves the liberation of embracing what

---

[23] On this theme, see Sallie McFague, *Life Abundant: Rethinking Theology and Economy for a Planet in Peril* (Minneapolis, MN: Fortress Press, 2001), especially 209–10.

truly matters to us, and knowing that all of it is absolute grace, the gift of a generous, loving God, and living into what matters.

Ecological Conversion: A Way of Praying

The Holy Spirit, I have been suggesting, can be seen as the Energy of Love at work in the whole of the natural world, the Breath of God, breathing through the whole of creation and through the lives and hearts of human beings. This creative Spirit is present in the natural world in countless ways that are far beyond the limits of the human. We can glimpse the wildness of the Spirit in the experiences we have of wilderness in nature. The uncontrollable Spirit of God is like the wild wind that "blows where it will" (John 3:8). But this Spirit is also boundless Love, closer to me than I am to myself. We encounter the Holy Spirit as the experience we have of the near- ness of God in mountains, deserts, forests, and seas, in the sense of being deeply connected with this place, this tree, this bird. The Spirit of God is Love that surrounds and sustains the uncounted insects, animals, and trees that I encounter. This Spirit is beyond the human and does not have a human face. Yet it is wonderfully personal Love that we meet at the heart of the natural world. The Holy Spirit is not less personal than human beings, but personal in a way that transcends the human way of being personal—a mys- terious Thou we encounter in our experience of the natural world.

We can be led, then, into a sense of mystery and presence in our experiences of the natural world that we can rightly name as the experience of God's Holy Spirit. Yet our experience of nature is, in itself, profoundly ambiguous. The natural world is unspeakably bountiful and beautiful but also a place of competition, violence, suffering, and death. In this ambiguity, it is the good news of God revealed to us in Jesus that is decisive for Christians. It is the Wisdom of God made flesh in Jesus that tells us that competition, suffering, and death are not the ultimate meaning of creation. In the Word made flesh, God is revealed as Love that embraces suffering creation, transforming it from within, bringing it to liberation and fulfillment.

It is only in the Word made flesh that we can know that in spite of all the violence and death, the glimpses we have of presence in the experience of the natural world can be trusted. It is only in the Word made flesh that we can dare to say, in the face of the ambiguity we find in the natural world, that Love is the meaning of the whole creation. Only then can I receive this tree before me, this bird outside my window, as giving expression in its own creaturely way to that same eternal Word of God that becomes flesh and has a human face in Jesus of Nazareth. In its unique, specific, and limited way, it is a revelatory word that speaks quietly and beautifully of the eternal Word and Wisdom of God.

The natural world, then, is a place of prayer, of encountering the mysterious presence of the Spirit, the divine Energy of Love at work in all the forms of life around us. It is the place for seeing this tree, this bird, this beach, as partaking of the Word of God, so that it is for me a word of that Word who is the eternal Wisdom of God. In the light of Christ I can see this creature before me as existing because it has been drawn into its own being and identity by the Divine Attractor, who draws the whole world of creatures into their existence within the community of creation and will draw all to their fulfillment.

As Athanasius made so clear, creation is a relationship of immediate presence of the triune God to each creature. This honeyeater sitting on the branch of a shrub two meters from me is not only the place of the Spirit's life-giving presence. It is not only a word of the Word to me, a presence of divine Wisdom. It is also gift of the Holy One who is Origin and Source of All, immediately present to this honeyeater through the Word and in the Spirit. When I am drawn to its uniqueness and beauty, I am drawn as well to that divine Wellspring of all creation, to the generative and creative Love that is at the heart of everything, to the One who is, for me and for the whole creation, like a father rushing to embrace a lost son or a mother comforting her child.

Index